T0211887

Simula SpringerBriefs on Computing

Reports on Computational Physiology

Volume 7

Editor-in-Chief

Aslak Tveito, Fornebu, Norway

Series Editors

Kimberly J. McCabe, Fornebu, Norway

Rachel Thomas, Fornebu, Norway

Andrew D. McCulloch, La Jolla, USA

In 2016, Springer and Simula launched an Open Access series called the Simula SpringerBriefs on Computing. This series aims to provide concise introductions to the research areas in which Simula specializes: scientific computing, software engineering, communication systems, machine learning and cybersecurity. These books are written for graduate students, researchers, professionals and others who are keenly interested in the science of computing, and each volume presents a compact, state-of-the-art disciplinary overview and raises essential critical questions in the field.

Simula's focus on computational physiology has grown considerably over the last decade, with the development of multi-scale mathematical models of excitable tissues (brain and heart) that are becoming increasingly complex and accurate. This sub-series represents a new branch of the SimulaSpringer Briefs that is specifically focused on computational physiology. Each volume in this series will explore multiple physiological questions and the models developed to address them. Each of the questions will, in turn, be packaged into a short report format that provides a succinct summary of the findings and, whenever possible, the software used will be made publicly available.

More information about this Subseries at http://www.springer.com/series/16669

Aslak Tveito · Kent-Andre Mardal · Marie E. Rognes

Editors

Modeling Excitable Tissue

The EMI Framework

 Springer

Editors
Aslak Tveito
Simula Research Laboratory
Fornebu, Norway

Marie E. Rognes
Simula Research Laboratory
Fornebu, Norway

Kent-Andre Mardal
Department of Mathematics
University of Oslo
Oslo, Norway

ISSN: 2512-1677   ISSN: 2512-1685   (electronic)
Simula SpringerBriefs on Computing
ISSN 2730-7735   ISSN 2730-7743   (electronic)
Reports on Computational Physiology
ISBN 978-3-030-61156-9   ISBN 978-3-030-61157-6   (eBook)
https://doi.org/10.1007/978-3-030-61157-6

Mathematics Subject Classification: 65M60, 65M06, 92C10, 92C37

This Springer imprint is published by the registered company Springer Nature Switzerland AG
The registered company address is: Gewerbestrasse 11, 6330 Cham, Switzerland

Foreword to *Reports on Computational Physiology*

Dear Reader,

In 2016, Springer and Simula launched an Open Access series called the Simula SpringerBriefs on Computing. This series aims to provide concise introductions to the research areas in which Simula specializes: scientific computing, software engineering, communication systems, machine learning and cybersecurity. These books are written for graduate students, researchers, professionals and others who are keenly interested in the science of computing. We know that entering a new field of research or getting up-to-date on a new topic of interest can be very demanding and time consuming, for students and experienced researchers alike. Each volume presents a compact, state-of-the-art disciplinary overview and raises essential critical questions in the field, all in approximately 100 pages.

Simula's focus on computational physiology has grown considerably over the last decade. Our researchers collaborate with partners around the world in interdisciplinary teams to develop multi-scale mathematical models of excitable tissues (brain and heart). These models are becoming increasingly complex and accurate, in particular as they are compared to experimental and clinical data. Since 2014, the University of California, San Diego (UCSD) and Simula have organized an annual summer school in computational physiology, in which graduate students spend the second two weeks of June in Oslo learning the principles underlying mathematical models commonly used to study the heart and the brain. The students are then assigned a research project to work on over the summer. In August the students travel to the University of California, San Diego to present their findings. Each year, we have been impressed by the students' abilities to learn the huge amount of mathematics and physiology theory required for their research projects, the results of which often contain the rudiments of a scientific paper.

As a result of our expanding activities in this field, we have decided to publish a branch of the SimulaSpringer Briefs that is specifically focused on computational physiology. Each volume in this series will explore multiple physiological questions and the models developed to address them. Each of the questions will, in turn, be packaged into a short report format (6-10 pages) that provides a succinct summary of the findings and, whenever possible, the software used will be made publicly available. All reports in this series are subjected to peer-review. We would like to emphasise that we do not require that reports represent new scientific results; rather, they can reproduce or supplement earlier computational studies or experimental findings.

The main driver for this series is to enable the publication of project reports from the annual summer school in Computational Physiology, but we will also publish related reports that fit the overall purpose of the SimulaSpringer Briefs. Due to the Covid-19 pandemic, the summer school in 2020 had to be cancelled, and as such this first issue of the new series is not a collection of project reports. However, the topic of the first issue fits very well in the framework of computational physiology; it presents novel methods and software for the simulation of excitable cells.

It is a pleasure to thank our collaborators at SpringerNature for their superbly efficient handling of this manuscript. In particular, we are grateful for the sound advice and support from Dr. Martin Peters, the Executive Editor for Mathematics, Computational Science and Engineering. We would also like to thank Dr. Henrik Nicolay Finsberg for his excellent technical support in this project.

Fornebu, Norway
September 2020

Dr. Kimberly J McCabe
Dr. Rachel Thomas
Dr. Andrew D McCulloch
Dr. Aslak Tveito

Preface

Partial differential equations (PDEs) have proved to be immensely useful in modelling Nature; virtually all fields of science have their own equations, and every field of engineering is based on mathematical models formulated in terms of PDEs. This is astonishing given the fact that no model, but the very simplest ones, can be studied using analytical (paper and pencil) techniques. Numerical computations have proved tremendously useful in order to understand models formulated in terms of PDEs, and it can be argued that the computer was invented for the purpose of solving such equations. The computer is extremely well suited to perform the huge amounts of tedious and highly repetitive computations that earlier had to be completed by humans. However, for a very long time, the computers typically available at research labs could solve only simple models. In the eighties, PDEs was almost always studied in 1 or 2 spatial dimensions; the 2D geometry was very simple and the model was most often linear and scalar. That level of computational complexity allowed analysis of qualitative properties of PDEs, but was insufficient for studying realistic models in Science and Engineering.

Over the past 30 years, we have witnessed a tremendous development in computing power both in terms of hardware, solution algorithms and software. This development has paved the way for realism in modeling; geometrical structures can now be represented with high degree of accuracy and complex systems of PDEs is applied to model the complex dynamics under consideration. This has led to greatly improved understanding throughout many branches of Science and had led to the development of considerably more accurate tools in Engineering.

Computational electrophysiology is a branch of Science that has benefited greatly from developments in computational analysis of PDEs. This development started 70 years ago with the celebrated paper by Hodgkin and Huxley (see (6)) and was followed 10 years later by the first cardiac action potential model developed by Noble (13). Since these groundbreaking papers, there have been intense efforts to understand how excitable cells works based on modeling and computations. This development has moved in tandem with new experimental techniques providing more accurate data necessary for parameterizing the increasingly complex models. A large number of membrane models (many hundreds) have been developed (see e.g. (11)), and these models have been used together with the monodomain or bidomain models (see e.g. (5; 21)) to study the electrochemical waves underpinning the contraction of the cardiac muscle. Similarly, the Cable equation (see e.g. (15)) has been extensively used to understand propagation of electrochemical signals by neurons.

Both the monodomain and the bidomain models of electrophysiology are derived based on homogenization of the cardiac tissue. In the resulting models, this means that the both the extracellular space, the cell membrane and the intracellular space are assumed to exist everywhere in the computational domain. Specifically, this means that the cardiac cell is not explicitly present in these models. This approach to modeling cardiac tissue enable analysis of phenomena on a relatively large length scale (mm), but is useless when it comes to study processes going on at a small length scale (μm). In 1993, the bidomain model was solved by Trayanova et al. ((20)) using 257 computational nodes, and at that time this was the best that technology would allow, and using a homogenized, large scale, model made perfect sense. More recently, however, models based on about 30 million mesh points are used allowing a characteristic mesh length of about 50μm which is about half the length of a human ventricular cell. Further refinement of the mesh used in the monodomain and bidomain models is not very useful since converged solutions are obtained at a quite coarse mesh (~0.3mm, see e.g. (12; 3)). This means that technology now allows simulation at a shorter length scale than the classical models (monodomain and bidomain) are meant to represent; it is impossible to gain information at the μm-level using these models. Specifically, the cell is impossible to represent explicitly in the classical models and that clearly limits their usefulness.

Interesting phenomena in electrophysiology take place close to the cell membrane. But since the cell is not explicitly present in the classical models, it is very difficult, if at all possible, to use such models to get a grip on what is going on in the vicinity of excitable cells. And since mesh resolution already has reached the μm-scale, it is clearly about time to introduce the cell as the building block in models of excitable tissue. In fact, this development started several years ago and has been pursued by many authors; see e.g. (7; 10; 14; 19; 16; 18; 17; 24; 1). Recently, we have followed up on these papers aiming at developing models, algorithms and software for cell-based representation of excitable tissue. We represent the extracellular (E) domain, the cell membrane (M) and the intracellular domain (I) explicitly, and therefore refer to it as the EMI-model.

In the first paper (see (23)) using the EMI model, we compared the results of the EMI model with the results of the classical Cable equations. In particular, we assessed the magnitude of the error introduced in the Cable equation by ignoring the ephaptic effects (i.e. assuming that the extracellular potential is constant). Then the same model was used to assess the effect on the action potential of placing a microelec-trode array in the extracellular domain (see (2)). Furthermore, we have developed computational techniques for solving the EMI equations (see (22; 8)) and used it to study properties of the conduction velocity of electrochemical wave traversing the cardiac muscle during a heart-beat (see (9)). Quite recently, the EMI model has been extended to also account for ion concentrations in the entire computational domain using the electroneutral Kirchhoff-Nernst-Planck (KNP) model, and the resulting model is referred to as the KNP-EMI model (see (4)). This opens the possibility for analysing the effect of depression waves traversing cortical tissue.

The main of advantage of the EMI-models is the possibility for the modeler to represent local properties of the cell and to study dynamics in the vicinity of the cells at the μm–scale. This opens vast possibilities for deeper understanding of the dynamics of collections of excitable cells. However, a main disadvantage is that the model is more complex and more computationally demanding that the common monodomain, bidomain, and cable equations. The purpose of this edition of the Simula SpringerBriefs on Computing is to provide succinct introductions to various aspects of the EMI-models, the solution algorithms and the software used to study these models. These models are not straightforward to implement and we therefore think it is useful to provide software for anyone interested in using the models. Note that it is specifically not our intention here to provide substantial new contributions to developments of models, algorithms or software, but rather to aid readers by providing easy and readable accounts of this material.

In order to avoid misunderstanding, we would like to add that we do not think this is the end of the monodomain-, bidomain-, or the cable-model. These models have been extremely useful and in combinations with membrane models they basically represent our collective knowledge about how excitable cells work. Much work remains to be done with these equations and the models we suggest are far too computationally demanding to be a realistic alternative for full scale simulations of human organs. Also, again in order to avoid misunderstandings, homogenization is still with us in the EMI models; the EMI models takes the typical length scale form mm to μm, but the atomic scale is still 10000 smaller, so the models we use in both E, M and I all represent averages.

Fornebu, Norway   *Aslak Tveito*
September 2020   *Kent-Andre Mardal*
                                                                                                Marie Rognes

References

1. Agudelo-Toro A, Neef A (2013) Computationally efficient simulation of electrical activity at cell membranes interacting with self-generated and externally imposed electric fields. Journal of Neural Engineering 10(2):026019
2. Buccino AP, Kuchta M, Jæger KH, Ness TV, Berthet P, Mardal KA, Cauwenberghs G, Tveito A (2019) How does the presence of neural probes affect extracellular potentials? Journal of neural engineering 16(2):026030
3. Costa CM, Neic A, Kerfoot E, Porter B, Sieniewicz B, Gould J, Sidhu B, Chen Z, Plank G, Rinaldi CA, Bishop MJ, Niederer SA (2019) Pacing in proximity to scar during cardiac resynchronization therapy increases local dispersion of repolarization and susceptibility to ventricular arrhythmogenesis. Heart Rhythm 16(10):1475 – 1483, DOI https://doi.org/10.1016/j.hrthm.2019.03.027, URL http://www.sciencedirect.com/science/article/pii/S1547527119302966, focus Issue: Sudden Death
4. Ellingsrud A, SolbråA, Einevoll G, Halnes G, Rognes ME (2020) Finite element simulation of ionic electrodiffusion in cellular geometries. Preprint
5. Franzone PC, Pavarino LF, Scacchi S (2014) Mathematical Cardiac Electrophysiology. Springer International Publishing
6. Hodgkin AL, Huxley AF (1952) A quantitative description of membrane current and its application to conduction and excitation in nerve. J Physiol 117:500–544
7. Hogues H, Leon LJ, Roberge FA (1992) A model study of electric field interactions between cardiac myocytes. IEEE Transactions on Biomedical Engineering 39(12):1232–1243
8. Jæger KH, Tveito A (2020) Efficient numerical solution of the EMI model representing extracellular space (E), the cell membrane (M) and the intracellular space (I) of a collection of cardiac cells. Preprint
9. Jæger KH, Edwards AG, McCulloch A, Tveito A (2019) Properties of cardiac conduction in a cell-based computational model. PLoS computational biology 15(5):e1007042
10. Krassowska W, Neu JC (1994) Response of a single cell to an external electric field. Biophysical Journal 66(6):1768–1776
11. Lloyd CM, Lawson JR, Hunter PJ, Nielsen PF (2008) The cellml model repository. Bioinformatics 24(18):2122–2123
12. Niederer S, Mitchell L, Smith N, Plank G (2011) Simulating human cardiac electrophysiology on clinical time-scales. Frontiers in Physiology 2(14):1–7
13. Noble D (1960) Cardiac action and pace-maker potentials based on the hodgkin-huxley equations. Nature 188:495–497
14. Roberts SF, Stinstra JG, Henriquez CS (2008) Effect of nonuniform interstitial space properties on impulse propagation: a discrete multidomain model. Biophysical Journal 95(8):3724–3737
15. Sterratt D, Graham B, Gillies A, Willshaw D (2011) Principles of computational modelling in neuroscience. Cambridge U. Press

16. Stinstra J, MacLeod R, Henriquez C (2010) Incorporating histology into a 3D microscopic computer model of myocardium to study propagation at a cellular level. Annals of Biomedical Engineering 38(4):1399–1414
17. Stinstra JG, Hopenfeld B, MacLeod RS (2005) On the passive cardiac conductivity. Annals of Biomedical Engineering 33(12):1743–1751
18. Stinstra JG, Roberts SF, Pormann JB, MacLeod RS, Henriquez CS (2006) A model of 3D propagation in discrete cardiac tissue. In: Computers in Cardiology, 2006, IEEE, pp 41–44
19. Stinstra JG, Henriquez CS, MacLeod RS (2009) Comparison of microscopic and bidomain models of anisotropic conduction. In: Computers in Cardiology, IEEE, pp 657–660
20. Trayanova NA, Roth BJ, Malden LJ (1993) The response of a spherical heart to a uniform electric field: a bidomain analysis of cardiac stimulation. IEEE transactions on biomedical engineering 40(9):899–908
21. Tung L (1978) A bidomain model for describing ischemic myocardial d-c potentials. PhD thesis, M.I.T. Cambridge, Mass.
22. Tveito A, Jæger KH, Kuchta M, Mardal KA, Rognes ME (2017) A cell-based framework for numerical modeling of electrical conduction in cardiac tissue. Frontiers in Physics 5:48
23. Tveito A, Jæger KH, Lines GT, Paszkowski Ł, Sundnes J, Edwards AG, Mäki-Marttunen T, Halnes G, Einevoll GT (2017) An evaluation of the accuracy of classical models for computing the membrane potential and extracellular potential for neurons. Frontiers in Computational Neuroscience 11:27
24. Ying W, Henriquez CS (2007) Hybrid finite element method for describing the electrical response of biological cells to applied fields. IEEE Transactions on Biomedical Engineering 54(4):611–620

List of Contributors

Aslak Tveito e-mail: aslak@simula.no
Karoline Horgmo Jæger e-mail: karolihj@simula.no
Åshild Telle e-mail: aashild@simula.no
Samuel T. Wall e-mail: samwall@simula.no
Joakim Sundnes e-mail: sundnes@simula.no
Ada J. Ellingsrud e-mail: ada@simula.no
Cécile Daversin-Catty e-mail: cecile@simula.no
Marie E. Rognes e-mail: meg@simula.no
Miroslav Kuchta e-mail: miroslav@simula.no
Jakob Schreiner e-mail: jakob@simula.no
Kent-André Mardal e-mail: kent_and@simula.no
Kristian G. Hustad e-mail: kghustad@simula.no
Xing Cai e-mail: xingca@simula.no
Henrik Nicolay Finsberg e-mail: henriknf@simula.no
Simula Research Laboratory, Martin Linges 25, 1364 Fornebu, Norway

Alessio Paolo Buccino e-mail: alessio.buccino@bsse.ethz.ch
Bio Engineering Laboratory, Department of Biosystems and Science Engineering,
ETH Zurich, Switzerland

Contents

Chapter 1

Derivation of a Cell-Based Mathematical Model of Excitable Cells

Karoline Horgmo Jæger[1] and Aslak Tveito[1,2]

Abstract Excitable cells are of vital importance in biology, and mathematical models have contributed significantly to understand their basic mechanisms. However, classical models of excitable cells are based on severe assumptions that may limit the accuracy of the simulation results. Here, we derive a more detailed approach to modeling that has recently been applied to study the electrical properties of both neurons and cardiomyocytes. The model is derived from first principles and opens up possibilities for studying detailed properties of excitable cells. We refer to the model as the EMI model because both the extracellular space (E), the cell membrane (M) and the intracellular space (I) are explicitly represented in the model, in contrast to classical spatial models of excitable cells. Later chapters of the present text will focus on numerical methods and software for solving the model. Also, in the next chapter, the model will be extended to account for ionic concentrations in the intracellular and extracellular spaces.

1.1 Introduction

Mathematical modeling has a great potential for increasing our understanding of the physiological processes underlying the function of the body. For example, modeling of the electrical properties of excitable cells may provide insight into the complex electrical signaling involved in a number of important functions, like transfer of information through neurons and coordination of the pumping of the heart. A popular model of the conduction of electrical signals in neurons is the so-called cable equation (30), whereas the extracellular potential surrounding neurons is often modeled using the point-source or line-source approximations (6). The three aforementioned models

---

[1] Simula Research Laboratory, Norway
[2] Department of Informatics, University of Oslo, Norway

The Author(s) 2021
A. Tveito et al. (eds.), *Modeling Excitable Tissue*, Simula SpringerBriefs on Computing 7, https://doi.org/10.1007/978-3-030-61157-6_1

have been used extensively to gain insight into the function of neurons and the interpretation of measurements of the extracellular potential around neurons (6; 5; 12). Correspondingly, the conduction of electrical signals through the heart is traditionally modeled using the homogenized bidomain and monodomain models (20). These models are also widely used and have, for instance, provided insight into mechanisms of cardiac arrhythmias (26; 33).

However, despite the success of the above-mentioned classical models of excitable cells, the models have certain shortcomings that may make them inaccurate or impractical in some situations. For example, in the derivation of the cable equation, the extracellular potential is often assumed to be constant (30; 15). Therefore, changes in the extracellular potential generated by the neuron itself or by neighboring neurons (i.e., ephaptic effects) are ignored in the model. This could potentially lead to inaccuracies (16; 2; 34). In addition, the point-source and line-source approximations rely on the assumption that the extracellular space is infinite and homogeneous. Consequently, the models might not be well-suited to interpret extracellular potentials measured when large measurement electrodes are present in the extracellular space close to the neurons (3).

Moreover, the bidomain and monodomain models represent cardiac tissue in a homogenized manner, assuming that the intracellular space, the extracellular space and the cell membrane exist everywhere in the tissue. Because the geometry of the individual cells is not represented, it is very hard to use the models to study the effect of, e.g., the cell geometry or a non-uniform distribution of ion channels on the cell membrane. These properties are both believed to influence cardiac conduction, but their exact effects are not fully understood and call for further investigations (28; 21). In addition, it has been proposed that ephaptic coupling between cardiac cells might occur at small extracellular clefts located at the intercalated discs between cells (29). Since the geometry of the extracellular space is not represented in the homogenized models, it is difficult to use these models to study such ephaptic effects.

In order to account for the difficulties related to the classical models, alternative electrophysiological models have been developed (e.g., (28; 31; 24)). In this chapter, we consider one of these alternative models, referred to as the EMI model, because it explicitly represents the extracellular space (E), the cell membrane (M) and the intracellular space (I). This model has been used to study both neurons (1; 34; 3) and cardiac tissue (32; 31; 18). Because the model represents the extracellular space, the membrane and the intracellular space in a coupled manner, the model allows for representation of ephaptic effects between neurons (34) or cardiomyocytes (18). In addition, since the geometry of the extracellular space is explicitly represented, the model allows for representation of non-homogeneous extracellular surroundings (3). Furthermore, since the geometry of each cell is represented, the model allows for study of the effect of cell geometry and non-uniform distributions of ion channels on cardiac conduction properties (18).

In other words, the EMI model allows for a more detailed representation of excitable cells and tissues than classical models of computational electrophysiology. In fact,

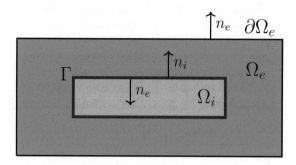

Fig. 1.1: Illustration of an EMI model domain consisting of an extracellular domain, Ω_e, a cell membrane, Γ, and an intracellular domain, Ω_i.

the classical models mentioned above can be derived from the more detailed EMI model by introducing certain simplifying assumptions, see e.g., (1; 8; 11). In this chapter, however, we focus on deriving the EMI model from Maxwell's equations of electromagnetism.

1.2 Derivation of the EMI Model

In this section, we present a derivation of the EMI model for excitable cells. This derivation is to a large extent based on the derivation found in (1; 17). We consider a domain separated into an extracellular part, Ω_e and an intracellular part, Ω_i, like illustrated in Figures 1.1 and 1.3. The cell membrane, denoted by Γ, is defined as the boundary between Ω_i and Ω_e. Here, we derive a model for the electrical potentials in both a domain with a single cell (Figure 1.1) and in a domain with two cells connected at an intercalated disc denoted by $\Gamma_{1,2}$ (Figure 1.3).

1.2.1 Fundamental Equations

We base the derivation of the EMI model on two of the quasi-static approximations of Maxwell's equations, i.e.,

$$\nabla \times \mathbf{E} = 0, \tag{1.1}$$
$$\nabla \times \mathbf{H} = \mathbf{J}. \tag{1.2}$$

Here, $\mathbf{E}$ is the electric field (typically in μF/cm), $\mathbf{H}$ is the magnetic field (typically in μA/cm) and $\mathbf{J}$ is the density of free currents (typically in μA/cm^2). In the quasi-static approximation of (1.2), it is assumed that free unbalanced charges are instantly balanced. The assumptions hold in the intracellular and extracellular spaces. However, we assume that charges may accumulate at the cell membrane and at the intercalated discs between cells. Therefore, we let (1.2) at these locations be replaced by the corresponding equation without the quasi-static approximation, i.e., by

$$\nabla \times \mathbf{H} = \mathbf{J} + \varepsilon \frac{\partial \mathbf{E}}{\partial t}, \tag{1.3}$$

where ε is the permittivity of the medium (typically in μF/cm). In addition, we assume that Ohm's law applies in the intracellular and extracellular domains. This means that

$$\mathbf{J} = \sigma \mathbf{E}, \tag{1.4}$$

where σ is the conductivity of the considered medium (typically in mS/cm). We also note that (1.1) implies that $\mathbf{E}$ is a conservative vector field and that it therefore can be defined as the gradient of a scalar field (9). More specifically, we can define

$$\mathbf{E} = -\nabla u, \tag{1.5}$$

where the scalar u is the electric potential (typically in mV).

1.2.2 Model for the Intracellular and Extracellular Domains

In order to derive equations for the intracellular and extracellular domains, we take the divergence of both sides of (1.2) and apply the vector identity $\nabla \cdot (\nabla \times \mathbf{H}) = 0$, which holds for any vector $\mathbf{H}$ (9). This yields

$$\nabla \cdot \mathbf{J} = 0.$$

Inserting (1.4) and (1.5), we obtain the Laplace equation

$$\nabla \cdot \sigma \nabla u = 0. \tag{1.6}$$

More specifically, for the intracellular and extracellular domains, we have

$$\nabla \cdot \sigma_i \nabla u_i = 0 \quad \text{in } \Omega_i, \tag{1.7}$$
$$\nabla \cdot \sigma_e \nabla u_e = 0 \quad \text{in } \Omega_e, \tag{1.8}$$

where σ_i and σ_e are the intracellular and extracellular conductivities, respectively, and u_i and u_e are the electric potentials in the intracellular and extracellular domains, respectively.

1.2.3 Model for the Membrane

In order to derive the EMI model equations for the membrane, we consider a volume element, B, intersected by the membrane. This volume element may be divided into an extracellular part, B_e, and an intracellular part, B_i, such that $B_e \cup B_i = B$ and $B_e \cap B_i = \emptyset$, as illustrated in Figure 1.2A. In each of these domains, we assume that (1.3) holds. Taking the divergence of both sides of (1.3) and applying the vector identity $\nabla \cdot (\nabla \times \mathbf{H}) = 0$ results in

$$\nabla \cdot \mathbf{J} = -\nabla \cdot \varepsilon \frac{\partial \mathbf{E}}{\partial t}. \tag{1.9}$$

Integrating this equation over each of the volume elements B_i and B_e, we get

$$\int_{B_i} \nabla \cdot \mathbf{J} \, dV = -\int_{B_i} \nabla \cdot \varepsilon \frac{\partial \mathbf{E}}{\partial t} \, dV,$$

$$\int_{B_e} \nabla \cdot \mathbf{J} \, dV = -\int_{B_e} \nabla \cdot \varepsilon \frac{\partial \mathbf{E}}{\partial t} \, dV,$$

and applying the divergence theorem (see e.g., (9)), we obtain

$$\int_{\partial B_i} \mathbf{J} \cdot \mathbf{n}_{B_i} \, dS = -\int_{\partial B_i} \varepsilon \frac{\partial \mathbf{E}}{\partial t} \cdot \mathbf{n}_{B_i} \, dS, \tag{1.10}$$

$$\int_{\partial B_e} \mathbf{J} \cdot \mathbf{n}_{B_e} \, dS = -\int_{\partial B_e} \varepsilon \frac{\partial \mathbf{E}}{\partial t} \cdot \mathbf{n}_{B_e} \, dS. \tag{1.11}$$

Here, $\mathbf{n}_{B_i}$ and $\mathbf{n}_{B_e}$ are the outward pointing normal vectors of B_i and B_e, respectively. Furthermore, in (1.10) and (1.11), the left-hand side terms represent the free ionic current and the right-hand side terms represent the capacitive current.

1.2.3.1 Ionic Current

We start by considering the left-hand side of (1.10), representing the ionic current. Here, we note that the boundary ∂B_i may be split into two parts, Γ_B and $\partial B_i \setminus \Gamma_B$, where Γ_B is the part of ∂B_i coinciding with the membrane and $\partial B_i \setminus \Gamma_B$ is the remaining part (see Figure 1.2A). We can then write

$$\int_{\partial B_i} \mathbf{J} \cdot \mathbf{n}_{B_i} \, dS = \int_{\partial B_i \setminus \Gamma_B} \mathbf{J} \cdot \mathbf{n}_{B_i} \, dS + \int_{\Gamma_B} \mathbf{J} \cdot \mathbf{n}_i \, dS, \tag{1.12}$$

where $\mathbf{n}_i$ is the outward pointing normal vector of the membrane and $\mathbf{n}_{B_i}$ is the outward pointing normal vector of the remaining part of B_i, as illustrated in Figure 1.2A. At the membrane, the current density, $\mathbf{J}$, consists of currents through different types of ion channels, pumps and exchangers located at the membrane. This current den-

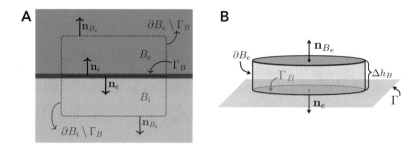

Fig. 1.2: **A**: Illustration of a volume element, B, intersected by the membrane. The volume element is separated into an intracellular part, B_i, and an extracellular part, B_e. **B**: Illustration of a small volume element, B, located on the extracellular part of the membrane.

sity is typically denoted by I_{ion} and given in units of $\mu A/cm^2$. By convention, I_{ion} is defined to be positive for a flux of positive ions out of the cell (i.e., in the direction of $\mathbf{n}_i$). This gives

$$\int_{\Gamma_B} \mathbf{J} \cdot \mathbf{n}_i \, dS = \int_{\Gamma_B} I_{\text{ion}} \, dS. \tag{1.13}$$

The boundary $\partial B_i \setminus \Gamma_B$ is located in the intracellular domain. Here, we assume that the current density, $\mathbf{J}$, is given by Ohm's law (1.4), such that

$$\int_{\partial B_i \setminus \Gamma_B} \mathbf{J} \cdot \mathbf{n}_{B_i} \, dS = \int_{\partial B_i \setminus \Gamma_B} \sigma_i \mathbf{E} \cdot \mathbf{n}_{B_i} \, dS. \tag{1.14}$$

Inserting (1.13) and (1.14) into (1.12), we get

$$\int_{\partial B_i} \mathbf{J} \cdot \mathbf{n}_{B_i} \, dS = \int_{\partial B_i \setminus \Gamma_B} \sigma_i \mathbf{E} \cdot \mathbf{n}_{B_i} \, dS + \int_{\Gamma_B} I_{\text{ion}} \, dS, \tag{1.15}$$

and similar arguments for the extracellular part of the membrane yield

$$\int_{\partial B_e} \mathbf{J} \cdot \mathbf{n}_{B_e} \, dS = \int_{\partial B_e \setminus \Gamma_B} \sigma_e \mathbf{E} \cdot \mathbf{n}_{B_e} \, dS - \int_{\Gamma_B} I_{\text{ion}} \, dS. \tag{1.16}$$

Note that the negative sign in front of the last term is due to the fact that $\mathbf{n}_e = -\mathbf{n}_i$.

1.2.3.2 Capacitive Current

For the right-hand side part of (1.10), representing the capacitive current, we again split the integral into two parts

$$\int_{\partial B_i} \varepsilon \frac{\partial \mathbf{E}}{\partial t} \cdot \mathbf{n}_{B_i} \, dS = \int_{\partial B_i \setminus \Gamma_B} \varepsilon \frac{\partial \mathbf{E}}{\partial t} \cdot \mathbf{n}_{B_i} \, dS + \int_{\Gamma_B} \varepsilon_\Gamma \frac{\partial \mathbf{E}}{\partial t} \cdot \mathbf{n}_i \, dS.$$

Here, ε_Γ is the permittivity of the membrane. Following the quasi-static assumptions, we assume that the term $\varepsilon \frac{\partial \mathbf{E}}{\partial t}$ is negligible for the part of ∂B_i that does not coincide with the membrane. Furthermore, from (1.5), we get $\mathbf{E} \cdot \mathbf{n}_i = -\nabla u \cdot \mathbf{n}_i \approx v/d$, where

$$v = u_i - u_e \tag{1.17}$$

is the membrane potential and d is the thickness of the membrane. We assume that the membrane can be treated as a capacitor formed by two parallel plates separated by an insulator. In that case, the membrane capacitance per area is given by $C_m = \varepsilon_\Gamma/d$ (13). Therefore,

$$\int_{\partial B_i} \varepsilon \frac{\partial \mathbf{E}}{\partial t} \cdot \mathbf{n}_{B_i} \, dS = \int_{\Gamma_B} \varepsilon_\Gamma \frac{\partial \mathbf{E}}{\partial t} \cdot \mathbf{n}_i \, dS = \int_{\Gamma_B} \frac{\varepsilon_\Gamma}{d} \frac{\partial v}{\partial t} \, dS = \int_{\Gamma_B} C_m \frac{\partial v}{\partial t} \, dS. \tag{1.18}$$

Similar arguments for the extracellular side yield

$$\int_{\partial B_e} \varepsilon \frac{\partial \mathbf{E}}{\partial t} \cdot \mathbf{n}_{B_e} \, dS = -\int_{\Gamma_B} C_m \frac{\partial v}{\partial t} \, dS, \tag{1.19}$$

where the change of sign again is due to the fact that $\mathbf{n}_e = -\mathbf{n}_i$.

1.2.3.3 Collecting the Ionic and Capacitive Currents

Collecting the ionic and capacitive currents by inserting (1.15)–(1.16) and (1.18)–(1.19) into (1.10)–(1.11), we obtain

$$\int_{\partial B_i \setminus \Gamma_B} \sigma_i \mathbf{E} \cdot \mathbf{n}_{B_i} \, dS + \int_{\Gamma_B} I_{\text{ion}} \, dS = -\int_{\Gamma_B} C_m \frac{\partial v}{\partial t} \, dS,$$

$$\int_{\partial B_e \setminus \Gamma_B} \sigma_e \mathbf{E} \cdot \mathbf{n}_{B_e} \, dS - \int_{\Gamma_B} I_{\text{ion}} \, dS = \int_{\Gamma_B} C_m \frac{\partial v}{\partial t} \, dS,$$

which can be rewritten to

$$\int_{\partial B_i \setminus \Gamma_B} \sigma_i \mathbf{E} \cdot \mathbf{n}_{B_i} \, dS = -\int_{\Gamma_B} I_m \, dS, \tag{1.20}$$

$$\int_{\partial B_e \setminus \Gamma_B} \sigma_e \mathbf{E} \cdot \mathbf{n}_{B_e} \, dS = \int_{\Gamma_B} I_m \, dS, \tag{1.21}$$

where the total membrane current density I_m is defined as

$$I_m = C_m \frac{\partial v}{\partial t} + I_{\text{ion}}. \tag{1.22}$$

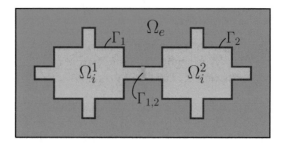

Fig. 1.3: Illustration of an EMI model domain consisting of two cells, Ω_i^1 and Ω_i^2, connected at an intercalated disc, $\Gamma_{1,2}$ and surrounded by an extracellular domain, Ω_e

We now wish to rewrite (1.20)–(1.21) to a differential form. We note that we can divide any volume element, B, intersecting the membrane into a purely intracellular, a purely extracellular, and a membrane intersecting part. We also know that (1.7)–(1.8) hold in the purely intracellular and extracellular parts. Therefore, we are interested in equations (1.20)–(1.21) as the size of B approaches zero. For example, we may consider a small extracellular volume element shaped as a cylinder, as illustrated in Figure 1.2B. As the height, Δh_B, of this cylinder approaches zero, the integral over $\partial B_e \setminus \Gamma_B$ approaches the integral over Γ_B, and we therefore get

$$\int_{\partial B_e \setminus \Gamma_B} \sigma_e \mathbf{E} \cdot \mathbf{n}_{B_e} \, dS \approx \int_{\Gamma_B} \sigma_e \mathbf{E} \cdot \mathbf{n}_{B_e} \, dS.$$

Inserting this approximation into (1.21), we obtain

$$\int_{\Gamma_B} \sigma_e \mathbf{E} \cdot \mathbf{n}_{B_e} \, dS = \int_{\Gamma_B} I_m \, dS \qquad \Rightarrow \qquad \sigma_e \mathbf{E} \cdot \mathbf{n}_{B_e} = I_m,$$

and inserting (1.5) and $\mathbf{n}_e = -\mathbf{n}_{B_e}$, we get

$$\sigma_e \nabla u_e \cdot \mathbf{n}_e = I_m. \tag{1.23}$$

Similar arguments for the intracellular part of the membrane yield

$$- \sigma_i \nabla u_i \cdot \mathbf{n}_i = I_m, \tag{1.24}$$

where the negative sign is due to the negative sign in (1.20). Finally, combining (1.23) and (1.24), we obtain

$$\sigma_e \nabla u_e \cdot \mathbf{n}_e = -\sigma_i \nabla u_i \cdot \mathbf{n}_i = I_m, \tag{1.25}$$

where $I_m = C_m \frac{\partial v}{\partial t} + I_{\text{ion}}$ and $v = u_i - u_e$ (see (1.22) and (1.17)).

1.2.4 Model for the Intercalated Disc

In some cases, we wish to model cells that are connected to each other, as illustrated in Figure 1.3. We then let the intercalated discs connecting the cells be represented as boundaries between the intracellular domains, like the membrane is a boundary between the intracellular and extracellular domains. Furthermore, we assume that the intercalated disc have capacitive properties like the membrane, and that gap junctions allow for currents between neighboring cells, in the same manner as ion channels allows for currents between the intracellular and extracellular spaces. Therefore, the derivation of equations for an intercalated disc follows the exact same lines as the derivation of the membrane equations. More precisely, for two connected cells, we define an intercalated disc potential, w, by

$$w = u_i^1 - u_i^2, \tag{1.26}$$

where u_i^1 and u_i^2 are the electric potentials in Ω_i^1 and Ω_i^2, respectively. In addition, we define a total intercalated disc current density, $I_{1,2}$, by

$$I_{1,2} = C_{1,2} \frac{\partial w}{\partial t} + I_{\text{gap}}, \tag{1.27}$$

where I_{gap} is the current density through the gap junctions, with positive direction in the direction from Ω_i^1 to Ω_i^2, $C_{1,2}$ is the capacitance of the intercalated disc, and $C_{1,2} \frac{\partial w}{\partial t}$ is the capacitive current density of the intercalated disc. Furthermore, following the same arguments as for the derivation of the membrane equations, we end up with an analogue to (1.25) of the form

$$\sigma_i^2 \nabla u_i^2 \cdot \mathbf{n}_i^2 = -\sigma_i^1 \nabla u_i^1 \cdot \mathbf{n}_i^1 = I_{1,2}, \tag{1.28}$$

representing the total current density across the interface.

1.2.5 Models of the Ionic Currents

Mathematical models of the ionic currents governing the membrane potential of excitable cells come in a large variety of versions; see (4) for several hundred examples. The simplest possible model is just a passive current of the form $I_{\text{ion}} = \text{const} \cdot v$, followed by a third order polynomial model. More realistic models tend to be more complex and are usually written on the form

$$I_{\text{ion}} = \sum_{i=1}^{N} I_i, \tag{1.29}$$

given in $\mu A/cm^2$. Here, the individual currents can usually be written on the form $I_i = I_i(v, s)$, where v denotes the membrane potential, given by $u_i - u_e$, and s denotes gating variables and ionic concentrations. The celebrated model of the action potential of a neuron presented by Hodgkin and Huxley (see (14)) can be written on this form, and so can the first model of a cardiac cell presented by Nobel (25). A comprehensive and readable introduction to models of the membrane ionic currents is given in the survey (27).

Correspondingly, the ionic currents through gap junctions between neighboring cells are often modeled by a simple passive model of the form $I_{gap} = \text{const} \cdot w$. More detailed models of voltage-dependent gap junction dynamics have also been introduced (see e.g., (10; 35)).

1.2.6 Summary of the Model Equations

In summary, the EMI model for a single cell surrounded by an extracellular domain (as illustrated in Figure 1.1) is given by the equations (1.7), (1.8), (1.17), (1.22) and (1.25), that is

$$\nabla \cdot \sigma_i \nabla u_i = 0 \qquad\qquad \text{in } \Omega_i, \qquad (1.30)$$

$$\nabla \cdot \sigma_e \nabla u_e = 0 \qquad\qquad \text{in } \Omega_e, \qquad (1.31)$$

$$\sigma_e \nabla u_e \cdot \mathbf{n}_e = -\sigma_i \nabla u_i \cdot \mathbf{n}_i \equiv I_m \quad \text{at } \Gamma, \qquad (1.32)$$

$$v = u_i - u_e \qquad\qquad \text{at } \Gamma, \qquad (1.33)$$

$$\frac{\partial v}{\partial t} = \frac{1}{C_m}(I_m - I_{\text{ion}}) \qquad \text{at } \Gamma, \qquad (1.34)$$

where u_i, u_e and v are the intracellular, extracellular and membrane potentials, respectively, typically given in mV. Moreover, σ_i and σ_e are the intracellular and extracellular conductivities, respectively (typically in mS/cm), C_m is the membrane capacitance (typically in $\mu F/cm^2$), and Γ denotes the cell membrane. The ionic currents through channels, pumps and exchangers at the membrane are denoted by I_{ion} and typically given in $\mu A/cm^2$.

If several cells are connected at intercalated discs, as illustrated for two cells in Figure 1.3, the system of equations must be extended to include equations for the currents between cells. For two cells, this extension consists of the equations

$$\sigma_i \nabla u_i^2 \cdot \mathbf{n}_i^2 = -\sigma_i \nabla u_i^1 \cdot \mathbf{n}_i^1 \equiv I_{1,2} \quad \text{at } \Gamma_{1,2}, \qquad (1.35)$$

$$u_i^1 - u_i^2 = w \qquad\qquad \text{at } \Gamma_{1,2}, \qquad (1.36)$$

$$w_t = \frac{1}{C_{1,2}}(I_{1,2} - I_{\text{gap}}) \qquad \text{at } \Gamma_{1,2}, \qquad (1.37)$$

where, as above, $\Gamma_{1,2}$ is the intercalated disc, $\mathbf{n}_i^1$ is the outward pointing normal vector of Ω_i^1, $\mathbf{n}_i^2$ is the outward pointing normal vector of Ω_i^2, and u_i^1 and u_i^2 are the intracellular potentials (typically in mV) of Ω_i^1 and Ω_i^2, respectively. Furthermore, $C_{1,2}$ is the specific capacitance of the intercalated disc (typically in $\mu F/cm^2$), and I_{gap} is the current through the gap junctions (typically in $\mu A/cm^2$).

1.3 Conclusion

In the present chapter, we have derived the EMI model. The EMI model predicts electrical potentials in cells with an explicit geometrical representation and thus allows for more detail than homogenized models of excitable tissue. In the next chapter (7, Chapter 2), the model will be extended by taking ion concentration in the extracellular and intracellular spaces into account. Numerical solutions of the EMI models will be presented in (19, Chapter 4), (23, Chapter 5) and (22, Chapter 6). In these chapters the readers will also be pointed to open software that can be used to solve the EMI model.

References

1. Agudelo-Toro A (2012) Numerical simulations on the biophysical foundations of the neuronal extracellular space. PhD thesis, Niedersächsische Staats-und Universitätsbibliothek Göttingen
2. Anastassiou CA, Perin R, Markram H, Koch C (2011) Ephaptic coupling of cortical neurons. Nature neuroscience 14(2):217–223
3. Buccino A, Kuchta M, Jæger KH, Ness T, Berthet P, Mardal KA, Cauwenberghs G, Tveito A (2019) How does the presence of neural probes affect extracellular potentials? Journal of Neural Engineering 16:026030
4. Cuellar AA, Lloyd CM, Nielsen PF, Bullivant DP, Nickerson DP, Hunter PJ (2003) An overview of CellML 1.1, a biological model description language. Simulation 79(12):740–747
5. Einevoll GT (2006) Mathematical modeling of neural activity. In: Dynamics of Complex Interconnected Systems: Networks and Bioprocesses, Springer, pp 127–145
6. Einevoll GT, Kayser C, Logothetis NK, Panzeri S (2013) Modelling and analysis of local field potentials for studying the function of cortical circuits. Nature Reviews Neuroscience 14(11):770
7. Ellingsrud AJ, Daversin-Catty C, Rognes ME (2020) A cell-based model for ionic electrod-iffusion in excitable tissue. In: Tveito A, Mardal KA, Rognes ME (eds) Modeling excitable tissue - The EMI framework, Simula Springer Notes in Computing, SpringerNature
8. Franzone PC, Pavarino LF, Scacchi S (2014) Mathematical cardiac electrophysiology, vol 13. Springer
9. Griffiths DJ (1989) Introduction to electrodynamics, 2nd edn. Prentice Hall
10. Henriquez AP, Vogel R, Muller-Borer BJ, Henriquez CS, Weingart R, Cascio WE (2001) In-fluence of dynamic gap junction resistance on impulse propagation in ventricular myocardium: a computer simulation study. Biophysical Journal 81(4):2112–2121
11. Henriquez CS, Ying W (2009) The bidomain model of cardiac tissue: from microscale to macroscale. In: Cardiac Bioelectric Therapy, Springer, pp 401–421
12. Herz AV, Gollisch T, Machens CK, Jaeger D (2006) Modeling single-neuron dynamics and computations: a balance of detail and abstraction. Science 314(5796):80–85
13. Hille B (2001) Ion channels of excitable membranes, vol 507. Sinauer Sunderland, MA
14. Hodgkin AL, Huxley AF (1952) A quantitative description of membrane current and its application to conduction and excitation in nerve. The Journal of Physiology 117(4):500–544
15. Holt GR (1998) A critical reexamination of some assumptions and implications of cable theory in neurobiology. PhD thesis, California Institute of Technology
16. Holt GR, Koch C (1999) Electrical interactions via the extracellular potential near cell bodies. Journal of computational neuroscience 6(2):169–184
17. Jæger KH (2019) Cell-based mathematical models of small collections of excitable cells. PhD thesis, University of Oslo

18. Jæger KH, Edwards AG, McCulloch A, Tveito A (2019) Properties of cardiac conduction in a cell-based computational model. PLoS computational biology 15(5):e1007042
19. Jæger KH, Hustad KG, Cai X, Tveito A (2020) Operator splitting and finite difference schemes for solving the emi model. In: Tveito A, Mardal KA, Rognes ME (eds) Modeling excitable tissue - The EMI framework, Simula Springer Notes in Computing, SpringerNature
20. Keener J, Sneyd J (2010) Mathematical physiology. Springer Science & Business Media
21. Kucera JP, Rohr S, Kleber AG (2017) Microstructure, cell-to-cell coupling, and ion currents as determinants of electrical propagation and arrhythmogenesis. Circulation: Arrhythmia and Electrophysiology 10(9):e004665
22. Kuchta M, Mardal KA (2020) Iterative solvers for cell-based emi models. In: Tveito A, Mardal KA, Rognes ME (eds) Modeling excitable tissue - The EMI framework, Simula Springer Notes in Computing, SpringerNature, pp 0–100
23. Kuchta M, Mardal KA, Rognes ME (2020) Solving the emi equations using finite element methods. In: Tveito A, Mardal KA, Rognes ME (eds) Modeling excitable tissue - The EMI framework, Simula Springer Notes in Computing, SpringerNature
24. Lin J, Keener JP (2010) Modeling electrical activity of myocardial cells incorporating the effects of ephaptic coupling. Proceedings of the National Academy of Sciences 107(49):20935–20940
25. Noble D (1962) A modification of the Hodgkin–Huxley equations applicable to purkinje fibre action and pacemaker potentials. The Journal of Physiology 160(2):317–352
26. Qu Z, Hu G, Garfinkel A, Weiss JN (2014) Nonlinear and stochastic dynamics in the heart. Physics Reports 543(2):61–162
27. Rudy Y (2012) From genes and molecules to organs and organisms: heart. Comprehensive Biophysics pp 268–327
28. Spach MS, Heidlage JF, Barr RC, Dolber PC (2004) Cell size and communication: role in structural and electrical development and remodeling of the heart. Heart Rhythm 1(4):500–515
29. Sperelakis N, McConnell K (2002) Electric field interactions between closely abutting excitable cells. IEEE Engineering in Medicine and Biology Magazine 21(1):77–89
30. Sterratt D, Graham B, Gillies A, Willshaw D (2011) Principles of computational modelling in neuroscience. Cambridge University Press
31. Stinstra JG, Roberts SF, Pormann JB, MacLeod RS, Henriquez CS (2006) A model of 3D propagation in discrete cardiac tissue. In: Computers in Cardiology, 2006, IEEE, pp 41–44
32. Stinstra JG, Henriquez CS, MacLeod RS (2009) Comparison of microscopic and bidomain models of anisotropic conduction. In: Computers in Cardiology, 2009, IEEE, pp 657–660
33. Trayanova NA (2011) Whole-heart modeling: applications to cardiac electrophysiology and electromechanics. Circulation Research 108(1):113–128
34. Tveito A, Jæger KH, Lines GT, Paszkowski Ł, Sundnes J, Edwards AG, Mäki-Marttunen T, Halnes G, Einevoll GT (2017) An evaluation of the accuracy of classical models for computing the membrane potential and extracellular potential for neurons. Frontiers in Computational Neuroscience 11:27
35. Weinberg SH (2017) Ephaptic coupling rescues conduction failure in weakly coupled cardiac tissue with voltage-gated gap junctions. Chaos: An Interdisciplinary Journal of Nonlinear Science 27(9):093908

Chapter 2

A Cell-Based Model for Ionic Electrodiffusion in Excitable Tissue

Ada J. Ellingsrud[1], Cécile Daversin-Catty[1] and Marie E. Rognes[1]

Abstract This chapter presents the KNP-EMI model describing ion concentrations and electrodiffusion in excitable tissue. The KNP-EMI model extends on the EMI model by removing the assumption that ion concentrations are constant in time and space, and may as such be more appropriate in connection with modelling e.g. spreading depression, stroke and epilepsy. The KNP-EMI model defines a system of time-dependent, nonlinear, mixed dimensional partial differential equations. We here detail the derivation of the system and present a numerical example illustrating how ion concentrations evolve during neuronal activity.

2.1 Introduction and Motivation

In this chapter, we present an extension of the EMI model, presented in (11, Chapter 1), describing ion concentrations and electrodiffusion in excitable tissue. The EMI model is based on the assumption that intra- and extracellular ion concentrations are constant in time and space. This is often a good approximation, as ion concentrations in healthy tissue typically quickly return to base levels after neuronal activity due to cellular mechanisms such as e.g. membrane pumps and glial cell buffering. However, there are scenarios where this assumption is inadequate.

Several cerebral pathologies are associated with increased neuronal activity (3), such as e.g. seizures and epilepsy (10; 6; 1), stroke (17), and spreading depression (22). In particular, periods of neuronal hyperactivity can lead to substantial variations in extracellular ion concentrations. These variations will in turn (i) influence membrane reversal potentials and (ii) generate diffusive currents. Changes in the reversal potentials, caused by local ionic shifts, may affect the dynamical properties of the

---

[1] Simula Research Laboratory, Norway

The Author(s) 2021
A. Tveito et al. (eds.), *Modeling Excitable Tissue*, Simula SpringerBriefs on Computing 7, https://doi.org/10.1007/978-3-030-61157-6_2

neurons (12; 16; 24). On the other hand, diffusive currents, driven by ion concentration gradients, can shift the extracellular potential (8; 3). Mathematical models addressing the aforementioned phenomena and pathologies should therefore also account for ion concentrations, their spatial and temporal gradients and associated dynamics.

In this chapter, we derive a system of time-dependent, nonlinear partial differential equations describing the distribution and evolution of ion concentrations in a geometrically-explicit representation of the intra- and extracellular domains using the electroneutral Kirchhoff-Nernst-Planck (KNP) model (21). We will refer to this model as the KNP-EMI model, see also e.g. (5).

2.2 Derivation of the Equations

Let the computational domain Ω and subdomains Ω_i, Ω_e, and Γ be defined as in the previous chapter 1.1. For simplicity and clarity, we present the mathematical model for one intracellular region $\Omega_{i1} = \Omega_i$ with membrane Γ below. We model a set K of intracellular and extracellular ion concentrations, and note that key ions in excitable tissue are potassium (K^+), sodium (Na^+), and chloride (Cl^-). For each ion species $k \in K$ and each region $r \in \{i, e\}$, we model the *ion concentrations* $c_r^k : \Omega_r \times (0, T] \rightarrow \mathbb{R}$ (mol/m^3), and *electrical potentials* $u_r : \Omega_r \times (0, T] \rightarrow \mathbb{R}$ (V), and additionally the *total transmembrane current density* $I_m : \Gamma \times (0, T] \rightarrow \mathbb{R}$ (A/m^2).

2.2.1 Equations in the Intracellular and Extracellular Volumes

In the EMI model, the free current densities $\mathbf{J}_i, \mathbf{J}_e$ (μA/cm^2), c.f. (1.4), are assumed to satisfy Ohm's law. To include diffusive ion effects, we instead assume that the free current density is composed of flux density contributions $\mathbf{J}_r^k$ (mol/(m^{2}s)) from different ions k as:

$$\mathbf{J}_r = \sum_{k \in K} F z^k \mathbf{J}_r^k \quad \text{in } \Omega_r, \tag{2.1}$$

where z^k is the valence of ion species k and F (C/mol) is Faraday's constant. Furthermore, we assume that ions can move by diffusion and/or in response to the electrical field as charged particles. Hence, the ion flux densities are modelled as the sum of two terms: (i) the ion concentrations that are transported via electrical potential gradients $\sigma_r^k \nabla u_r$ and (ii) the diffusive movement of ions due to ionic gradients $D_r^k \nabla c_r^k$:

$$\mathbf{J}_r^k = -\sigma_r^k \nabla u_r - D_r^k \nabla c_r^k \quad \text{in } \Omega_r, \tag{2.2}$$

where D_r^k (m²/s) and σ_r^k denote the effective diffusion coefficient and the conductivity for ion species k in region r, respectively. The conductivity σ_r^k depends on the concentration of ion species k and the diffusion coefficient D_r^k in the following manner:

$$\sigma_r^k = \sigma_r^k(c_r^k) = \frac{D_r^k z^k}{\psi} c_r^k \quad \text{in } \Omega_r. \tag{2.3}$$

Here, the constant $\psi = RTF^{-1}$ combines Faraday's constant F, the absolute temperature T (K), and the gas constant R (J/(K mol)). Moreover, the bulk conductivity σ_r can be expressed as:

$$\sigma_r = \sigma_r(c_r^k) = \frac{F}{\psi} \sum_{k \in K} D_r^k c_r^k (z^k)^2 \quad \text{in } \Omega_r. \tag{2.4}$$

See e.g. (21) for a derivation of the conductivity (2.3) and the bulk conductivity (2.4). Comparing with (1.4) and (1.5), we note the dependency on the ion concentrations in the conductivity σ_r in (2.3), and the second term accounting for ion diffusion in (2.2).

As in Chapter 1, we stipulate that:

$$\nabla \cdot \mathbf{J}_i = 0 \quad \text{in } \Omega_i, \tag{2.5}$$

$$\nabla \cdot \mathbf{J}_e = 0 \quad \text{in } \Omega_e. \tag{2.6}$$

Finally, conservation of ions for the bulk of each region Ω_r gives that:

$$\frac{\partial [k]_i}{\partial t} + \nabla \cdot \mathbf{J}_i^k = 0 \quad \text{in } \Omega_i, \tag{2.7}$$

$$\frac{\partial [k]_e}{\partial t} + \nabla \cdot \mathbf{J}_e^k = 0 \quad \text{in } \Omega_e, \tag{2.8}$$

for $t \in (0,T]$.

2.2.2 Membrane Currents

We next turn to modelling the cell membrane currents and membrane potential across the interface Γ. As in Chapter 1, we introduce the membrane potential v as the jump in the electrical potential over the membrane:

$$v = u_i - u_e \quad \text{on } \Gamma. \tag{2.9}$$

We also introduce the total membrane current as the combination of a capacitive current and ion specific currents:

$$I_m = I_{\text{cap}} + I_{\text{ion}} = C_m \frac{\partial v}{\partial t} + I_{\text{ion}}, \tag{2.10}$$

where the total channel current I_{ion} is the sum of the ion specific channel currents I_{ion}^k:

$$I_{\text{ion}} = \sum_{k \in K} I_{\text{ion}}^k, \quad I_{\text{ion}}^k = I_{\text{ion}}^k(v, c_{\cdot}^k, \ldots). \tag{2.11}$$

The channel currents I_{ion}^k are subject to modelling, and will be discussed briefly in Section 2.2.2.1.

Using our concepts, we have that the *total ionic current density* $I_m : \Gamma \times (0, T) \to \mathbb{R}$ (A/m^2) across the interface Γ (from the intracellular to the extracellular domain) is given by:

$$-F \sum_{k \in K} z^k \mathbf{J}_e^k \cdot \mathbf{n}_e = F \sum_{k \in K} z^k \mathbf{J}_i^k \cdot \mathbf{n}_i \equiv I_m. \tag{2.12}$$

It now remains to specify a set of interface conditions for the specific ion fluxes $\mathbf{J}_r^k \cdot \mathbf{n}_r$ for $r \in \{i, e\}$.

Here, we propose a heuristic approach via ion specific capacitive current modelling, and note that an alternative approach is presented in (15). As for the total current, we assume that the capacitive current can be represented as a sum of ion specific contributions:

$$I_{\text{cap}} = \sum_{k \in K} I_{\text{cap}}^k. \tag{2.13}$$

Without loss of generality, we let the ion specific capacitive current $I_{\text{cap},r}^k$ in region Ω_r at the interface Γ be some fraction α_r^k of the total capacitive current I_{cap}:

$$I_{\text{cap},r}^k = \alpha_r^k I_{\text{cap}}. \tag{2.14}$$

Specifically, we assume that:

$$\alpha_r^k = \frac{D_r^k (z^k)^2 [k]_r}{\sum_{l \in K} D_r^l (z^l)^2 [l]_r}, \tag{2.15}$$

and note that $\sum_{k \in K} \alpha_r^k = 1$ for $r \in \{i, e\}$. By the above definitions, (2.10) and (2.12), we let the intracellular and extracellular ion fluxes across the membrane be given by:

$$\mathbf{J}_i^k \cdot \mathbf{n}_i = \frac{I_{\text{ion}}^k + \alpha_i^k (I_m - I_{\text{ion}})}{F z^k}, \quad -\mathbf{J}_e^k \cdot \mathbf{n}_e = \frac{I_{\text{ion}}^k + \alpha_e^k (I_m - I_{\text{ion}})}{F z^k}, \tag{2.16}$$

for $k \in K$.

2.2.2.1 Modelling Specific Ion Channels

The membrane channel currents $I_{\text{ion}}^k(v)$ for each ion species k are subject to modelling. These currents are typically expressed on the form:

$$I_{\text{ion}}^k(v) = g_L^k(v - E^k), \qquad (2.17)$$

where g_L^k is the conductivity, and E^k is the ion specific reversal potential (or Nernst potential), given by:

$$E^k = \frac{RT}{z^k F} \ln \frac{c_e^k}{c_i^k}. \qquad (2.18)$$

This Nernst potential depends on the concentration ratio, whereas the Nernst potential in models without explicit modelling of ion concentrations is constant. Typical models include synaptic input currents, passive neuronal leak channels, or e.g. the Hodgkin-Huxley model (9). For more details on membrane current models and modelling, see e.g. (18).

2.2.3 Summary of KNP-EMI Equations

The KNP-EMI model equations follow from inserting (2.1) into (2.5)–(2.6), combined with (2.7), (2.8), (2.9), (2.10), and (2.16), and read as follows.

For each ion species $k \in K$ and each region $r \in \{i, e\}$, find the *ion concentrations* $c_r^k : \Omega_r \times (0, T] \to \mathbb{R}$ (mol/m³), the *electrical potentials* $u_r : \Omega_r \times (0, T] \to \mathbb{R}$ (V), and the *total transmembrane current density* $I_m : \Gamma \times (0, T] \to \mathbb{R}$ (A/m²) such that[1]:

$$\nabla \cdot \left(F \sum_k z^k \mathbf{J}_r^k \right) = 0 \qquad \text{in } \Omega_r, \qquad (2.19)$$

$$\frac{\partial c_r^k}{\partial t} + \nabla \cdot \mathbf{J}_r^k = 0 \qquad \text{in } \Omega_r, \qquad (2.20)$$

$$-F \sum_k z^k \mathbf{J}_e^k \cdot \mathbf{n}_e = F \sum_k z^k \mathbf{J}_i^k \cdot \mathbf{n}_i \equiv I_m \qquad \text{at } \Gamma, \qquad (2.21)$$

$$v = u_i - u_e \qquad \text{at } \Gamma, \qquad (2.22)$$

$$\frac{\partial v}{\partial t} = \frac{1}{C_m}(I_m - I_{\text{ion}}) \qquad \text{at } \Gamma, \qquad (2.23)$$

where the ion flux density $\mathbf{J}_r^k$ is given by (2.2), and I_{ion} is subject to modelling. A set of initial and boundary or compatibility conditions will close the system.

---

[1] Note that the additional negative signs in (2.19) and (2.21), compared with the corresponding equations in Chapter 1, result from our physically consistent definition of the ion flux density $\mathbf{J}_r^k$ as the negative gradient, cf. (2.2).

2.3 Numerical Solution of the KNP-EMI Equations

The KNP-EMI model defines a complicated system of time-dependent, nonlinear, mixed dimensional partial differential equations. The number of unknowns depends on the number of ion species modelled. Some of the variables exist in the intracellular and extracellular domains, while others live on the lower-dimensional membrane. This setting is numerically challenging and calls for advanced techniques.

To solve the KNP-EMI model numerically, one may consider a finite difference scheme to approximate the time derivatives, a linearization of ion flux densities $\mathbf{J}_r^k$ and fractions α_r^k, a splitting scheme to handle active ion channel current models, and a finite element discretization in space. Such a solution algorithm is detailed in (5), and we refer the reader to this description for further details.

2.4 Comparing KNP-EMI and EMI during Neuronal Hyperactivity

Neurons are negatively charged relative to their environment, with a resting membrane potential of about -70 mV. This resting potential is maintained by low concentrations of sodium ions (Na^+) and high levels of potassium ions (K^+) inside the cell (23). Action potentials (neuronal activity) are generated by the opening of sodium and potassium channels in the cell membranes. The ionic gradient will drive sodium into the cell and depolarize the cell membrane. Next, the potassium channels open causing an outflux of potassium which in turn repolarizes the cell.

As a result, there is a continuous need to pump potassium into the intracellular space and sodium out to the extracellular space to restore the electrochemical gradient across the cell membrane. One of the key mechanisms for this process is the Na/K/ATPase pump. The Na/K/ATPase pump actively transports 3 Na^+ ions out of the cell and 2 K^+ ions into the cell (7; 14; 20). Several pathologies are associated with increased neuronal activity, e.g. seizures and epilepsy (10; 6; 1), and spreading depression (22). In periods of neuronal hyperactivity, the Na/K/ATPase pumps may not be able to restore the concentrations to baseline levels. Consequently, the electrochemical gradients may be reduced, and silenced neuronal activity and cellular swelling may occur (13).

The ion concentration gradients observed during neuronal hyperactivity thus yields a suitable setting for illustrating differences between the KNP-EMI and the EMI frameworks. In particular, we compare the two frameworks both during normal neuronal activity (firing rate of 1 Hz) and during hyperactivity (firing rate of 50 Hz).

2.4.1 Model Parameters and Membrane Mechanisms

We consider two idealized axons, represented by two parallel, rectangular domains, surrounded by extracellular space in three dimensions. The diameter of each axon is $2.0 \cdot 10^{-7}$ m, and they are separated by $1.0 \cdot 10^{-7}$ m of extracellular space. Parameter values are as listed in Table 2.1. We refer to the supplementary code for a complete description of the model set-up (4).

KNP-EMI membrane mechanisms The membrane mechanisms in the KNP-EMI model, cf. (2.11), are modelled using the standard Hodgkin-Huxley model (9) combined with a model for the Na/K/ATPase pump (12), the KCC2 cotransporter (24) and the NKCC1 cotransporter (24). The Na/K/ATPase pump current I_{ATP} (A/m^2) is modelled as:

$$I_{\text{ATP}} = I_{\text{ATP}}(c_i^{\text{Na}}, c_e^{\text{K}}) = \frac{\hat{I}}{(1 + \frac{m_{\text{K}}}{c_e^{\text{K}}})^2(1 + \frac{m_{\text{Na}}}{c_i^{\text{Na}}})^3}, \tag{2.24}$$

where $\hat{I}$ is the maximum pump strength and m_{K} and m_{Na} denote the pump threshold for extracellular potassium and intracellular sodium, respectively. Further, the transmembrane currents generated by the KCC2 cotransporter I_{KCC2} (A/m^2) and the NKCC1 cotransporter I_{NKCC1} (A/m^2) are modelled as:

$$I_{\text{KCC2}} = S_{\text{KCC2}} \ln(\frac{c_i^{\text{K}} c_i^{\text{Cl}}}{c_e^{\text{K}} c_e^{\text{Cl}}}), \tag{2.25}$$

$$I_{\text{NKCC1}} = S_{\text{NKCC1}} \frac{1}{1 + e^{16 - c_e^{\text{K}}}} (\ln(\frac{c_i^{\text{K}} c_i^{\text{Cl}}}{c_e^{\text{K}} c_e^{\text{Cl}}}) + \ln(\frac{c_i^{\text{Na}} c_i^{\text{Cl}}}{c_e^{\text{Na}} c_e^{\text{Cl}}})), \tag{2.26}$$

where S_{KCC2} and S_{NKCC1} are the maximal cotransporter strengths. Moreover, the cell is stimulated by prescribing a synaptic input I_{syn} of the form:

$$I_{\text{syn}}^k = g_{\text{syn}} H e^{\frac{t - t_0}{\alpha}} (v - E^k), \tag{2.27}$$

where α (s) is the synaptic time constant, H is the Heaviside function for a small region on the left side of the axons, and $g_{\text{syn}} = 1.25 \cdot 10^{-3}$ S/m^2. In summary, the membrane channel currents for sodium, potassium and chloride are modelled as:

$$I_{\text{ion}}^{\text{Na}}(v, c_r^k) = g_{\text{leak}}^{\text{Na}}(v - E^{\text{Na}}) + \bar{g}^{\text{Na}} m^3 h(v - E^{\text{Na}}) + 3I_{\text{ATP}} + I_{\text{NKCC1}} + I_{\text{syn}}^{\text{Na}}$$

$$I_{\text{ion}}^{\text{K}}(v, c_r^k) = g_{\text{leak}}^{\text{K}}(v - E^{\text{K}}) + \bar{g}^{\text{K}} n^4(v - E^{\text{K}}) - 2I_{\text{ATP}} + I_{\text{NKCC1}} + I_{\text{KCC2}}$$

$$I_{\text{ion}}^{\text{Cl}}(v, c_r^k) = g_{\text{leak}}^{\text{Cl}}(v - E^{\text{Cl}}) - 2I_{\text{NKCC1}} - I_{\text{KCC2}},$$

where, g_{leak}^k and $\bar{g}^k$ is the leak conductivity and the maximal conductivity for ion species k, respectively, the Nernst potential E^k for ion species k is as described in Section 2.2.2.1, and the gating variables m, h and n are described by the standard Hodgkin-Huxley ODEs, see e.g. (23) for details.

EMI membrane mechanisms For the EMI model, we apply the standard Hodgkin-Huxley model and stimulate the cell by prescribing an input current of the form (2.27); thus, the membrane channels currents are modelled as:

$$I_{\text{ion}}(v) = g_{\text{leak}}^{\text{Na}}(v - E^{\text{Na}}) + g_{\text{leak}}^{\text{K}}(v - E^{\text{K}}) + g_{\text{leak}}^{\text{Cl}}(v - E^{\text{Cl}})$$
$$+ \bar{g}^{\text{Na}}m^3 h(v - E^{\text{Na}}) + \bar{g}^{\text{K}}n^4(v - E^{\text{K}}) + I_{\text{ATP}} + I_{\text{syn}},$$

where E^{K}, E^{Na} and E^{Cl} are calculated by (2.18) with the initial values from the KNP-EMI model for the sodium and potassium concentrations. Similarly, the bulk conductivities σ_i and σ_e are calculated by (2.4), and the net current from the Na/K/ATPase pump I_{ATP} is given by (2.24). Finally, there is no contribution from KCC2 and NKCC1, as both cotransporters mediate ion transport without any net charge movement across the membrane.

Parameter	Symbol	Value	Unit	Reference
gas constant	R	8.314	J/(K mol)	(23)
temperature	T	300	K	(23)
Faraday's constant	F	$9.648 \cdot 10^4$	C/mol	(23)
membrane capacitance	C_m	0.02	F/m	(24)
Na$^+$ diffusion coefficient	D_r^{Na}	$1.33 \cdot 10^{-9}$	m^2/s	(23)
K$^+$ diffusion coefficient	D_r^{K}	$1.96 \cdot 10^{-9}$	m^2/s	(23)
Cl$^-$ diffusion coefficient	D_r^{Cl}	$2.03 \cdot 10^{-9}$	m^2/s	(23)
intracellular immobile anions	c_i^{A}	110	mM	
extracellular immobile anions	c_e^{A}	10	mM	
valence of immobile anions	z_A	-1		
Na$^+$ leak conductivity	g_L^{Na}	0.281	S/m^2	*
K$^+$ leak conductivity	g_L^{K}	0.43	S/m^2	*
Cl$^-$ leak conductivity	g_L^{Cl}	0.2	S/m^2	*
K$^+$ HH max conductivity	$\bar{g}^{\text{K}}$	360	S/m^2	(9)
Na$^+$ HH max conductivity	$\bar{g}^{\text{Na}}$	1200	S/m^2	(9)
maximum pump strength	$\hat{I}$	0.18	A/m^2	(24)
maximum KCC2 strength	S_{KCC2}	0.0034	A/m^2	*
maximum NKCC1 strength	S_{NKCC1}	0.023	A/m^2	*
ECS K$^+$ pump threshold	m_{K^+}	3	mM	*
ICS Na$^+$ pump threshold	m_{Na^+}	12	mM	*
synaptic time constant	α	$1.0 \cdot 10^{-3}$	s	
global time step	Δt	$1.0 \cdot 10^{-5}$	s	
local time step	Δt^*	$\Delta t/25$	s	
spatial resolution	$\Delta x = \Delta y$	$2.5 \cdot 10^{-7}$	m	

Table 2.1: The physical and model parameters used in the simulations. The values are collected from Sterratt et al. (23), Hodgkin et al. (9), Wei et al. (24), whereas the values marked with * are computed by a steady state estimation.

The initial conditions for the intra- and extracellular ion concentrations, the membrane potential and the gating variables are listed in Table 2.2. At the exterior boundary, we apply no flux boundary conditions for each ion species.

Parameter	Symbol	Value	Unit	Reference
initial intracellular Na$^+$ concentration	$c_i^{Na,0}$	18	mM	
initial extracellular Na$^+$ concentration	$c_e^{Na,0}$	120	mM	
initial intracellular K$^+$ concentration	$c_i^{K,0}$	80	mM	
initial extracellular K$^+$ concentration	$c_e^{k,0}$	4	mM	
initial intracellular Cl$^+$ concentration	$c_i^{Cl,0}$	7	mM	
initial extracellular Cl$^+$ concentration	$c_e^{Cl,0}$	112	mM	
initial membrane potential	v^0	$-67.74 \cdot 10^{-3}$	V	*
initial HH gating value (Na$^+$ activation)	m^0	$\frac{\alpha_m(v^0)}{\alpha_m(v^0)+\beta_m(v^0)}$	–	(9)
initial HH gating value (Na$^+$ inactivation	h^0	$\frac{\alpha_h(v^0)}{\alpha_h(v^0)+\beta_h(v^0)}$	–	(9)
initial HH gating value (K$^+$ activation)	n^0	$\frac{\alpha_n(v^0)}{\alpha_n(v^0)+\beta_n(v^0)}$	–	(9)

Table 2.2: Initial conditions. The initial ion concentrations are chosen such that the Nernst potentials are equal to those in the Hodgkin-Huxley model (9). The membrane potential is computed by a steady state estimation.

2.4.2 Results and Discussion

During normal activity, the KNP-EMI and the EMI models behave similarly, both for the membrane potential and the extracellular potential (Figure 2.1 A, B). The stimuli current depolarizes the membrane potential above the threshold for firing, and an action potential is initiated (Figure 2.1 A). Simultaneously, the extracellular potential decreases by ~ 0.13 mV, before quickly returning to baseline (Figure 2.1 B).

During hyperactivity, the KNP-EMI and EMI models differ (Figure 2.1 C, D, E, F). In both models, repeated action potentials are triggered. But, for the KNP-EMI model, we observe changes in the membrane potential between hyperpolarization phases. In particular, we conclude that the KNP-EMI membrane resting potential increases with repeated firing: after 5 action potentials (at $t = 90$ ms) the membrane potential has a minimum value of -75 mV, which is an 9% increase from the first action potential. Eventually, the membrane is depolarized to the point where action potentials can long longer be fired (Figure 2.1 E).

The observed changes are caused by alterations in the ion concentration gradients. For each action potential, the extracellular Na$^+$ concentration decreases by 0.15 mM and the extracellular K^+ concentration increases by 0.16 mM (Figure 2.2 A, B). During normal activity (Figure 2.2 A, B), the ion concentrations will slowly be pumped back toward baseline levels, and the membrane potentials are not substantially affected by the small ion concentration changes. However, in the case of hyperactivity, the membrane mechanisms (i.e. pumps and cotransporters) are not able to keep up. Consequently, the extracellular Na^+ concentration will keep decreasing and the extracellular potassium will keep increasing, causing the cell to depolarize (Figure 2.2 C, D).

In the KNP-EMI model (Figure 2.2 A, B), we note that 7.92 % of the extracellular K^+ concentration is restored, and 7.3 % of the extracellular Na^+ concentration is restored after 100 ms. That is, the extracellular concentrations do not reach baseline levels within the simulation period. Other studies have reported that it takes on the order of minutes (0.5 minutes (19), 6 minutes (2)) before the concentrations return to baseline after neuronal activity.

2.5 Conclusions and Outlook

In this chapter, we have presented a mathematical model, the KNP-EMI model, for ionic electrodiffusion in excitable tissue with an explicit representation of the intracellular, extracellular and membrane domains. For further reading on methodological aspects, we refer to (5; 15) and references therein. This model extends on the EMI model presented in Chapter 1 and may be more accurate in situations with rapid and persistent changes in ion concentrations. Moreover, the KNP-EMI framework allows for modelling ligand-gated ion channels (e.g. NMDA receptors).

The complexity of the KNP-EMI system yields a number of numerical challenges. The mere number of unknowns result in large systems of equations calling for efficient solution techniques. The nonlinearities in the system can easily lead to non-convergence and thus call for robust algorithms. Moreover, the coupling of full and lower dimensional domains and fields calls for well-posed numerical methods together with suitable simulation software. Further, the system couples different time scales: from neuronal action potentials taking place at the microscale to the slower diffusion process. In short, modelling ionic electrodiffusion in the EMI setting is an area with vast opportunities for further research.

Acknowledgements The authors would like to thank Geir Halnes for useful discussions and Min Ragan-Kelley for technical assistance.

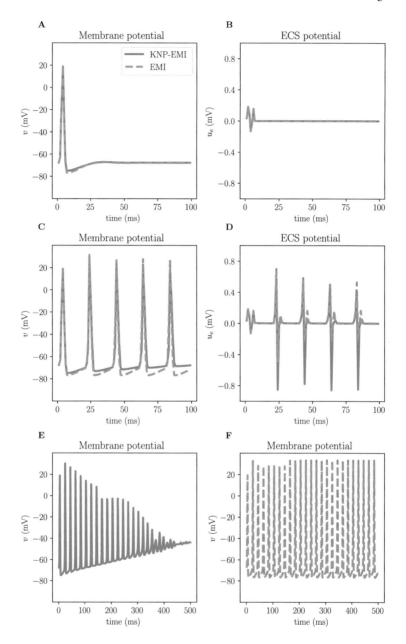

Fig. 2.1: Comparison of potentials over time at fixed points in space predicted by the KNP-EMI and the EMI frameworks during normal activity (upper panels) and during hyperactivity (mid and lower panels). The membrane potentials for KNP-EMI and EMI during normal activity (**A**) and hyperactivity (**C**, **E**, **F**), and the extracellular potentials for KNP-EMI and EMI during normal activity (**B**) and hyperactivity (**D**).

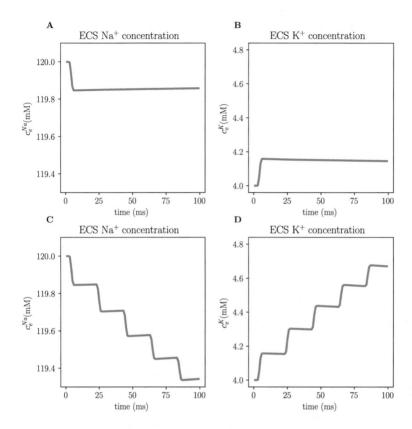

Fig. 2.2: Time development of extracellular ion concentrations at a fixed point in space for the KNP-EMI framework during normal activity (upper panels) and hyperactivity (lower panels). The extracellular sodium (**A**) and potassium (**B**) concentrations during normal activity, and the extracellular sodium (**C**) and potassium (**D**) concentrations during hyperactivity.

References

1. Bragin A, Engel Jr J, Wilson CL, Fried I, Mathern GW (1999) Hippocampal and entorhinal cortex high-frequency oscillations (100–500 Hz) in human epileptic brain and in kainic acid-treated rats with chronic seizures. Epilepsia 40(2):127–137
2. Chander BS, Chakravarthy VS (2012) A computational model of neuro-glio-vascular loop interactions. PloS one 7(11)
3. Dietzel I, Heinemann U, Lux H (1989) Relations between slow extracellular potential changes, glial potassium buffering, and electrolyte and cellular volume changes during neuronal hyperactivity in cat brain. Glia 2(1):25–44
4. Ellingsrud AJ (2020) Supplementary material (code) for Chapter 2 in 'EMI: Cell based mathematical model of excitable cells' (version 2.0). DOI 10.5281/zenodo.3767058, URL `http://doi.org/10.5281/zenodo.3767058`
5. Ellingsrud AJ, Solbrå A, Einevoll GT, Halnes G, Rognes ME (2020) Finite element simulation of ionic electrodiffusion in cellular geometries. Frontiers in Neuroinformatics 14:11
6. Fisher RS, Webber W, Lesser RP, Arroyo S, Uematsu S (1992) High-frequency EEG activity at the start of seizures. Journal of clinical neurophysiology: official publication of the American Electroencephalographic Society 9(3):441–448
7. Gadsby DC (2009) Ion channels versus ion pumps: the principal difference, in principle. Nature reviews Molecular cell biology 10(5):344–352
8. Halnes G, Mäki-Marttunen T, Keller D, Pettersen KH, Andreassen OA, Einevoll GT (2016) Effect of ionic diffusion on extracellular potentials in neural tissue. PLoS computational biology 12(11):e1005193
9. Hodgkin AL, Huxley AF (1952) A quantitative description of membrane current and its application to conduction and excitation in nerve. The Journal of physiology 117(4):500–544
10. Jacobs J, LeVan P, Chander R, Hall J, Dubeau F, Gotman J (2008) Interictal high-frequency oscillations (80–500 Hz) are an indicator of seizure onset areas independent of spikes in the human epileptic brain. Epilepsia 49(11):1893–1907
11. Jæger KH, Tveito A (2020) Derivation of a cell-based mathematical model of excitable cells. In: Tveito A, Mardal KA, Rognes ME (eds) Modeling excitable tissue - The EMI framework, Simula Springer Notes in Computing, SpringerNature
12. Kager H, Wadman WJ, Somjen GG (2000) Simulated seizures and spreading depression in a neuron model incorporating interstitial space and ion concentrations. Journal of neurophysiology 84(1):495–512
13. Kempski O (2001) Cerebral edema. In: Seminars in nephrology, Elsevier, vol 21, pp 303–307
14. de Lores Arnaiz GR, Ordieres MGL (2014) Brain Na+, K+-ATPase activity in aging and disease. International journal of biomedical science: IJBS 10(2):85

15. Mori Y, Peskin C (2009) A numerical method for cellular electrophysiology based on the electrodiffusion equations with internal boundary conditions at membranes. Communications in Applied Mathematics and Computational Science 4(1):85–134
16. Øyehaug L, Østby I, Lloyd CM, Omholt SW, Einevoll GT (2012) Dependence of spontaneous neuronal firing and depolarisation block on astroglial membrane transport mechanisms. Journal of computational neuroscience 32(1):147–165
17. Rabiller G, He JW, Nishijima Y, Wong A, Liu J (2015) Perturbation of brain oscillations after ischemic stroke: a potential biomarker for post-stroke function and therapy. International journal of molecular sciences 16(10):25605–25640
18. Rudy Y (2012) From genes and molecules to organs and organisms: heart. Comprehensive Biophysics pp 268–327
19. Sætra MJ, Einevoll GT, Halnes G (2020) An electrodiffusive, ion conserving Pinsky-Rinzel model with homeostatic mechanisms. bioRxiv
20. Scheiner-Bobis G (2002) The sodium pump. European Journal of Biochemistry 269(10):2424–2433
21. Solbrå A, Wigdahl BA, van den Brink Jonas, Anders MS, T EG, Geir H (2018) A Kirchhoff-Nernst-Planck framework for modeling large scale extracellular electrodiffusion surrounding morphologically detailed neurons. PLOS Computational Biology 14(10):1–26, DOI 10.1371/journal.pcbi.1006510, URL https://doi.org/10.1371/journal.pcbi.1006510
22. Somjen GG (2001) Mechanisms of spreading depression and hypoxic spreading depression-like depolarization. Physiological reviews 81(3):1065–1096
23. Sterratt D, Graham B, Gillies A, Willshaw D (2011) Principles of computational modelling in neuroscience. Cambridge University Press
24. Wei Y, Ullah G, Schiff SJ (2014) Unification of neuronal spikes, seizures, and spreading depression. Journal of Neuroscience 34(35):11733–11743

Chapter 3

Modeling Cardiac Mechanics on a Sub-Cellular Scale

Åshild Telle[1], Samuel T. Wall[1] and Joakim Sundnes[1]

Abstract We aim to extend existing models of single-cell mechanics to the EMI framework, to define spatially resolved mechanical models of cardiac myocytes embedded in a passive extracellular space. The models introduced here will be pure mechanics models employing fairly simple constitutive laws for active and passive mechanics. Future extensions of the models may include a coupling to the electrophysiology and electro-diffusion models described in the other chapters, to study the impact of spatially heterogeneous ion concentrations on the cell and tissue mechanics.

3.1 Introduction

A vast range of models have been developed for the force development of cardiac and skeletal muscle, on the scale of a single cross bridge (10), myofilament (3), sarcomere (2), and the complete cell (13). The scales involved and the main functional units considered on each scale are schematically illustrated in Figure 3.1. Common to most existing models is the fact that they focus on a single spatial scale, and any coupling between scales is fairly crudely represented. As an example, the model by Rice et al. (13) is essentially a model of a single sarcomere (Fig. 3.1 D), which is normalized and then scaled to yield a realistic force output for cell- and tissue-level mechanics applications. Other models provide detailed descriptions of mechanisms and interactions on a molecular level (Fig. 3.1 F)(4; 3), and are able to capture many of the characteristic non-linearities of muscle cell mechanics. However, key aspects of mechanical activation and force-length relationships are still not fully understood, and they may be the result of interactions between individual sarcomeres and myofibril bundles. A few attempts have been made at modeling interactions at this scale, and

---

[1] Simula Research Laboratory, Norway

The Author(s) 2021
A. Tveito et al. (eds.), *Modeling Excitable Tissue*, Simula SpringerBriefs on Computing 7, https://doi.org/10.1007/978-3-030-61157-6_3

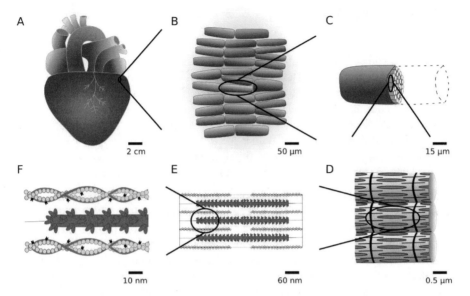

Fig. 3.1: The heart (A) is mainly composed of cardiac muscle cells, also called cardiomyocytes (B). Each cell (C) is composed of long tubes of sarcomeres (D), in which the thin and thick myofilaments overlap in layers (E). The interaction between these (F) causes the cardiac muscle to contract in a process called the cross-bridge cycle.

have shown potentially interesting emergent behaviours (2; 11). Furthermore, heart failure and other pathologies are linked with heterogeneous intracellular calcium concentration resulting from disruptions of the calcium regulation system. Describing the effect of such heterogeneities on the cell contraction and force development requires spatially resolved mechanics models on the sub-cellular scale.

Finite element models of contracting myocytes have been proposed (8; 14), and have been used to explore the impact of model assumptions, calcium heterogeneity, and boundary conditions. The model presented by Ruiz-Baier et al. (14) describes the individual myocyte as a hyperelastic material, and uses an active strain approach to describe the contraction. Both the passive and active mechanical properties are assumed to be homogeneous, but sub-cellular heterogeneities can easily be introduced. We here propose to extend the single myocyte model in (14) to include the extracellular domain, and to model collections of cells, based on similar ideas used for the electrophysiology model presented in (17; 18) and (7, Chapter 1).

3.2 Models and Methods

The motion and deformation of the heart can be described by the classical theory of non-linear solid mechanics. The primary unknown in our computational model will be the displacement vector u, which for each *material point* describes the difference between its current and original position. We have $u = x - X$, where X is the original (reference) position of a point, and x is its position after the deformation. From the displacement vector we can define the *deformation gradient* $F = \partial x/\partial X = I + \partial u/\partial X$, which is an essential quantity describing the deformation of a solid. See for instance (6) for a detailed introduction to non-linear solid mechanics.

A characteristic feature of the heart and other muscles is that they contract and deform even in the absence of external loads. The overall deformation and mechanical state of the heart depends both on this active contraction and on the passive mechanical properties of the tissue. There are two main approaches for modeling the coupling of active and passive mechanics in cardiac tissue, often referred to as *active strain* and *active stress*. Both approaches are based on modeling the active and passive contributions separately, then combining them into a complete coupled model.

In the active strain approach, the active-passive coupling is incorporated through a multiplicative decomposition of the deformation gradient F into active and passive components, $F = F_p F_a$. Here, F_a represents an active deformation governed by the cell state, and F_p is a passive elastic deformation which ensures compatibility with loads and kinematic boundary conditions. The active stress approach is based on an additive split of the stress tensor into its active and passive components. In terms of the first Piola-Kirchhoff stress tensor P, the stress is written as $P = P_p + P_a$, where P_a is a function of the cellular activation state and P_p is a standard elastic stress derived from a strain energy function.

Both of these approaches have their strengths and weaknesses. In general, the active strain approach is considered to be more suitable for deriving mathematically well-behaved constitutive laws, while the active stress concept is more easily coupled to biophysically detailed models of cell contraction.

3.2.1 Fundamental Equations

In this study we will primarily use the active stress approach, but for completeness we also present the equations arising from the active strain approach. This model can be derived as a direct extension of the single myocyte model in (14), using a similar approach as in (17; 18) to consider both the intra- and extracellular domains:

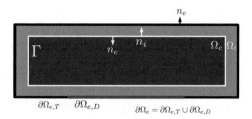

Fig. 3.2: Illustration of the intra- and extracellular domains for a single cell and its surroundings.

$$
\begin{aligned}
&a:\ \nabla \cdot P_i = 0, &\quad b:\ P_i = \frac{\partial \Psi_i}{\partial F_i}, &\quad c:\ F_i = F_i^p F_i^a, &\quad \text{in } \Omega_{\mathrm{i}}, \\
&d:\ \nabla \cdot P_e = 0, &\quad e:\ P_e = \frac{\partial \Psi_e}{\partial F_e}, & &\quad \text{in } \Omega_{\mathrm{e}}, \\
&f:\ u_i = u_e, &\quad g:\ n_i \cdot P_i = n_e \cdot P_e, & &\quad \text{on } \Gamma, &\quad (3.1) \\
&h:\ n_e \cdot P_e = 0, & & &\quad \text{on } \partial \Omega_{\mathrm{e,T}}, \\
&i:\ u = 0, & & &\quad \text{on } \partial \Omega_{\mathrm{e,D}}.
\end{aligned}
$$

Here, Ω_i and Ω_e are the intra- and extracellular domains, respectively, Γ is the interface between the domains, with the normal vector n_i pointing out of the intracellular domain and n_e out of the extracellular domain. Finally, $\partial \Omega_{e,T}$ and $\partial \Omega_{e,D}$ are the parts of the outer boundary $\partial \Omega_e$ subject to traction- and displacement boundary conditions, respectively. See Figure 3.2 for a sketch of a typical computational domain, including a single cell and its immediate surroundings. Following (14), we here apply the active strain approach to incorporate active contraction of the myocyte, where the intracellular deformation gradient F_i is decomposed as described above. The passive part is assumed to be hyper-elastic and derived from a strain energy function, see for instance (6) for details. A common choice for the active part is $F_i^a = diag((1 - \gamma), (1 - \gamma)^{-1/2}, (1 - \gamma)^{-1/2})$, where γ describes the fiber contraction and is a function of the cell activation state. For a more detailed introduction and discussion of active strain models, we refer to (1).

The active stress model is the most widely used approach for modeling coupled active and passive mechanics on tissue level, and this is the approach we will employ in the subsequent numerical experiments. In the present context the active stress model involves a decomposition of the intracellular first Piola-Kirchhoff stress P_i into a passive elastic part P_i^p and an active part P_i^a. The passive stress is derived from a strain energy function in the usual way, while the active stress is a function of the cell activation state. For the simplified model considered here we write the active stress as a function of time and the local fiber stretch λ, but the approach can easily be extended to include detailed biophysical models of the contractile mechanisms. The full active stress model may be written as

$$
\begin{aligned}
&a: \nabla \cdot P_i = 0, && b: P_i = \frac{\partial \Psi_i}{\partial F} + P_i^a(t, \lambda), && \text{in } \Omega_i, \\
&c: \nabla \cdot P_e = 0, && d: P_e = \frac{\partial \Psi_e}{\partial F_e}, && \text{in } \Omega_e, \\
&e: u_i = u_e, && f: n_i \cdot P_i = n_e \cdot P_e, && \text{on } \Gamma, \\
&g: n_e \cdot P_e = 0 && && \text{on } \partial \Omega_{e,T}, \\
&h: u = 0 && && \text{on } \partial \Omega_{e,D}.
\end{aligned}
\tag{3.2}
$$

Both approaches treat the extracellular domain in the same way, as a passive hyperelastic material governed by a strain energy function Ψ_e. As given by (3.1) f-g and (3.2) g-h we assume continuity of stresses P_i, P_e and displacements u_i, u_e across the cell membrane Γ, implying that the membrane itself has no stiffness. The outer boundary Ω_e is assumed to be stress free, with Dirichlet conditions applied to parts of the boundary to avoid rigid body motion. Models for the active stress P_i^a come in many forms, including simple phenomenological models as well as detailed biophysical models of cell electro-mechanics (12; 13). For the present study we apply a simple model where the active stress is derived from a (pseudo-) strain energy in the same way as the passive stress:

$$
P_i^a = \frac{\partial \Psi_i^a}{\partial F}.
\tag{3.3}
$$

Here, Ψ_i^a is given by

$$
\Psi_i^a = \frac{T_{active}(t)}{2} \lambda^2,
$$

where $\lambda = ||Fe_1||$ is the stretch in the so-called *fiber direction* (i.e. the main orientation of the muscle cells), defined by the unit vector e_1, and $T_{active}(t)$ is a prescribed function defining the active contractile force as a function of time.

3.2.2 Specific Model Choices

In this section we describe specific choices of the constitutive laws describing active and passive material properties in the models above, to arrive at a complete model that can be solved for the deformations and stresses. As noted above, we will in the following only consider the active stress model, given by (3.2). For the strain energy defining the passive stress-strain relationships we have applied a model from (19), which belongs to the family of models first presented by Guccione et al. (5). The same form of strain energy is used in the intra- and extracellular domains, but we allow the material parameters to be different. Both domains are modeled as nearly incompressible, with volume changes during deformations controlled by a penalty term. We have

$$
\Psi_i = C_i(e^{Q_i} - 1) + \kappa(J \ln J - J + 1) \qquad x \in \Omega_i,
\tag{3.4}
$$

$$
\Psi_e = C_e(e^{Q_e} - 1) + \kappa(J \ln J - J + 1) \qquad x \in \Omega_e,
\tag{3.5}
$$

where Q_i, Q_e are functions depending on components of the Green-Lagrange strain tensor $E = \frac{1}{2}(F^T F - I)$:

$$Q_j = b_{f,j}E_{11}^2 + b_{t,j}(E_{22}^2 + E_{33}^2 + E_{23}^2 + E_{32}^2)$$
$$+ b_{fs,j}(E_{12}^2 + E_{21}^2 + E_{13}^2 + E_{31}^2). \tag{3.6}$$

Furthermore $C_j, b_{f,j}, b_{t,j}$, and $b_{fs,j}$, for $j = i, e$ are material parameters characterizing the material's stiffness to the various strain modes, κ is a penalty parameter that controls the volume changes, and $J = \det F$. For a fully incompressible deformation we have $J = 1$, and in our nearly incompressible model we tune the parameter κ to keep $J \approx 1$.

In its most general form, the materials described by (3.4)-(3.6) are are transversely isotropic, which is a special case of orhtotropic materials. While an orthotropic material has different mechanical properties in three different directions, a transversely isotropic material is isotropic in planes normal to a characteristic direction. Passive cardiac tissue is known to behave as an orthortopic material (9), with the three directions dictated by the orientation and organization of the myocytes. However, a transversely isotropic material is shown to be a good approximation, with material isotropy in planes normal to the fiber direction, the main orientation of the muscle cells. The details of the intra- and extracellular material behavior in our micro-structural model are less well-studied, and the degree of anisotropy has not been characterized. From the microstructure of the contractile apparatus occupying most of the intracellular space (see Figure 3.1) it is natural to assume anisotropic behavior, but the exact degree of aniostropy is not known. As a starting point, we set the intracellular material parameters to

$$b_{f,i} = 8, \qquad b_{t,e} = 2, \qquad b_{fs,e} = 4. \tag{3.7}$$

For the extracellular space we assume isotropic material behaviour, setting

$$b_{f,e} = b_{t,e} = b_{fs,e} = 1. \tag{3.8}$$

The bulk compressibility was set to $\kappa = 1000\,\text{kPa}$ in both domains, while we explored different values of the scaling parameters C_i and C_e, to be specified below.

For the active stress model defined in (3.3) we have used a pre-computed transient tension $T_{active}(t)$ as shown in Figure 3.3. The curve was computed using the model of Rice et al. (13) with default parameters, which outputs a normalized force. This value was then scaled such that the peak value reaches 2 kPa, giving a reasonable contractile stress for our application.

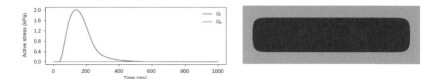

Fig. 3.3: Transient tension $T_{active}(t)$ over time (left), first computed in (13), then scaled to give values on a reasonable scale. In the intracellular domain the active tension is homogeneously set to this value; in the extracellular domain there is no such tension, implemented as being set to zero for all time steps.

3.2.3 Numerical Methods

The problem defined by (3.2) is solved with the displacement u as the primary unknown. To solve the system with the finite element method, it is convenient to formulate it as a single PDE defined over the entire domain $\Omega = \Omega_i \cup \Omega_e$. Such a formulation is not possible for the strong form of the PDEs, so we first need to derive the weak form of the equations. Starting with (3.2)a, we define a suitable vector function space $V(\Omega_i)$ defined over the intracellular domain, multiply the equation with a test function $v \in V(\Omega_i)$ and integrate by parts, to arrive at a weak formulation

$$\int_{\Omega_i} P_i \cdot \nabla v\, dx - \int_\Gamma (n_i \cdot P_i) v = 0.$$

This equation is to be satisfied for all $v \in V(\Omega_i)$. Performing the same steps for the extracellular domain, and using the boundary condition (3.2)g on the outer boundary, we get

$$\int_{\Omega_e} P_e \cdot \nabla v\, dx - \int_\Gamma (n_e \cdot P_e) v = 0.$$

This equation should hold for all test functions $v \in V(\Omega_e)$, where $V(\Omega_e)$ is a suitable space of functions defined over the domain Ω_e. Using similar arguments as in (15), we can define a function space $V(\Omega)$ as the set of functions defined over Ω that belong to both $V(\Omega_i)$ and $V(\Omega_e)$ and are continuous over Γ. With this definition, we may add the two weak forms above to obtain

$$\int_{\Omega_i} P_i \cdot \nabla v\, dx - \int_\Gamma (n_i \cdot P_i) v + \int_{\Omega_e} P_e \cdot \nabla v\, dx - \int_\Gamma (n_e \cdot P_e) v = 0.$$

Which is to be satisfied for all $v \in V(\Omega)$. Since $n_e = -n_i$, the surface integrals over Γ cancel because of (3.2)f. We can also use (3.2)e to define a single displacement field over Ω, and we are left with the following weak form: Find $u \in V(\Omega)$ such that

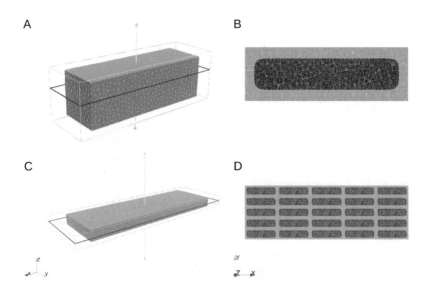

Fig. 3.4: **A**: Volume element of one single cell; lines indicate cross section area. **B**: Cross section along longitudial direction of the cell. **C**: Volume element, 5 x 5 cells; lines indicate cross section area. **D**: Cross section along longitudial direction of the cells.

$$\int_{\Omega} P \cdot \nabla v dx = 0, \tag{3.9}$$

is satisfied for all $v \in V(\Omega)$. with P defined by (3.2)b and (3.2)d in the respective domains.

3.3 Results

We here present a number of numerical experiments to illustrate the general behavior of the models defined above. The code is implemented using FEniCS, and an archieved version of the code is available, see (16).

For the simulations we used two different meshes; one representing a single cell and one representing a sheet of five by five cells, see Figure 3.4. Both meshes include subdomains defining the intra- and extracellular domains, separated by the cell membrane. To avoid rigid body motion, we keep a few points in the middle fixed. The rest of the boundary is kept unloaded to allow free contraction of the cells.

For each experiment we calculated the Green-Lagrange strain tensor E and the Cauchy stress tensor σ, given by

$$\sigma = \begin{cases} |F|^{-1} P_i F^T & x \in \Omega_i \\ |F|^{-1} P_e F^T & \text{otherwise.} \end{cases}$$

On matrix form we can can write these out as

$$E = \begin{bmatrix} E_{11} & E_{12} & E_{13} \\ E_{21} & E_{22} & E_{23} \\ E_{31} & E_{32} & E_{33} \end{bmatrix} \qquad \sigma = \begin{bmatrix} \sigma_{11} & \sigma_{12} & \sigma_{13} \\ \sigma_{21} & \sigma_{22} & \sigma_{23} \\ \sigma_{31} & \sigma_{32} & \sigma_{33} \end{bmatrix}$$

and for each of these we present plots for the first and the middle components, $(E_{11}, E_{22}, \sigma_{11}, \sigma_{22})$, which characterize strain and stress in the fiber and cross-fiber directions.

Fig. 3.5: Tracking points, for which we evaluate functions of interest across various experiments. The points are uniformly distributed on a line from one corner to the middle, in the xy-direction, corresponding to the cross-section shown in Figure 3.4. Two of them are both located in the extracellular subdomain, and one should expect them to show different patterns than the three located in the intracellular subdomain.

We first considered a single cell, and simulated contraction over a single cardiac cycle with homogeneous active force applied throughout the cell. For this simulation we chose parameter values $C_e = C_i = 0.5$. The results are presented in Figure 3.6, where we observe that the deformation follows the expected pattern of a contraction in the longitudinal direction of the cell. Furthermore, in spite of the homogeneous applied active stress we see slight spatial variations in the deformation state, resulting from the discontinuity of active force across the cell membrane.

Similar patterns are observed in the simulation of the sheet of 25 cells, shown in Figure 3.7. In this experiment the same active stress transient through the intracellular domain of all the cells, with the same material parameters. We still observe spatial variations in the deformation pattern – each cells is affected by mechanical deformation around them.

We then considered two cases where we kept all parameters but one fixed, exploring the choices of material stiffness parameters C_e and C_i. The results are presnted in

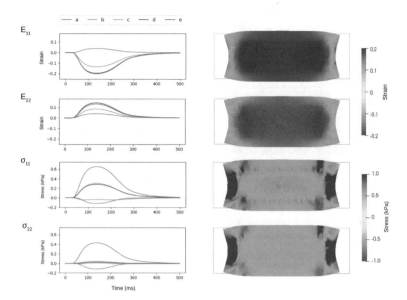

Fig. 3.6: First and middle components of the Cauchy stress tensor σ and Green-Lagrange strain E, for a single cell. The plots to the left shows values plotted over time, for the first 500 ms (out of 1000), following tracking points as shown in Figure 3.5. The plots to the right shows values over a cross-section as shown in Figure 3.4, as the active tension reaches it's peak value. The grey rectangle indicates initial configuration.

Figures 3.8 and 3.9. These simulations were again performed on a mesh representing a single cell, with active force applied as described above. For the first experiment we kept C_e fixed at 0.5, changing C_i; that is, we let the material stiffness in the extra-cellular domain remain the same while increasing the stress/strain scaling parameter in the intracellular domain. As C_i increases the material becomes stiffer, and for the same active stress applied, one should expect less contraction. This can indeed be observed; both components of the Cauchy stress tensor (in magnitude) and the strain tensor decreases everywhere, and for the last three parameter choices there is almost no difference in deformation. On the other hand, we still apply an active stress in the intracellular domain, and we observe that the strain close to the membrane doesn't change much even if it changes everywhere else.

For the next experiment we changed to keeping $C_i = 0.5$ constant, while increasing C_e. We can observe higher Cauchy stress for the first component, and lower Cauchy stress for the second component, with increasing values of C_e. The strain decreases for both components. This is exactly as expected – in one end of the spectrum, having

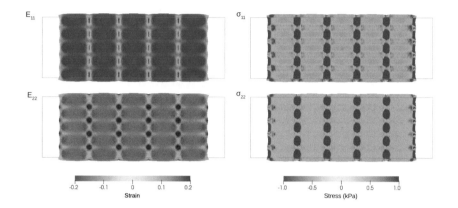

Fig. 3.7: First and middle components of the Cauchy stress tensor σ, and and Green-Lagrange strain E, for 5 x 5 cells. Values are plotted over the cross-section as shown in 3.4, as the active tension reaches it's peak value. The grey rectangle indicates initial configuration.

$C_e = 0.5$, one would expect the extracellular subdomain to not affect the intracellular domain as it's rather flexible. For a given tension in the intracellular domain, it will just move along quite easily, while the overall behaviour in the whole domain is governed by the contraction inside the cell. As C_e increases, the material is modeled as stiffer and hence constrain the movement more. For very high values the material is so stiff that it hardly moves, efficiently keeping the membrane close to fixed.

3.4 Discussion

We have presented a general framework for modeling cardiac mechanics on a sub-cellular scale, by extending a model of the type defined in (14) to the extracellular domain. A series of preliminary numerical experiments demonstrate that the model behaves as expected, with the discontinuity across the cell membrane giving rise to spatially varying deformation fields even though both the active stress and other model parameters are spatially homogeneous over the intracellular domain.

The main purpose of this work was to present the model framework and to illustrate the general behaviour of the model, while more detailed investigations and model extensions are left for future studies. A complete list of model limitations and potential extensions would be too extensive to present here, but it is worth commenting on a few of the most obvious ones. First, the model derivation above included a number of simplifying assumptions on the mechanical properties of the cell membrane and

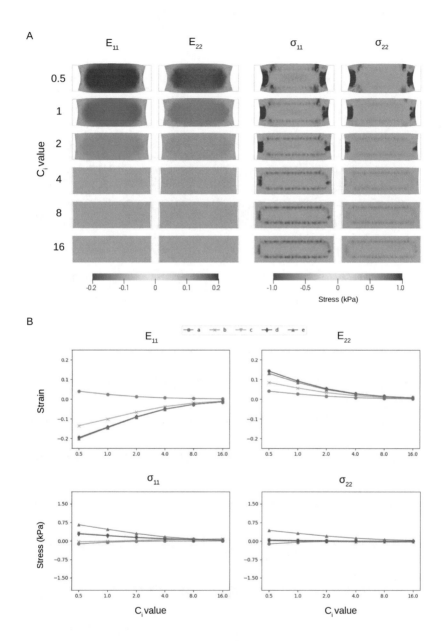

Fig. 3.8: First and middle components of the Cauchy stress tensor σ and Green-Lagrange strain E, for a single cell, as we vary the parameter C_i, which defines the stiffness of the material in the intracellular domain. Panel A shows spatial variation over a cross-section of the cell (see Figure 3.4), at peak. Panel B shows how the value, at peak, changes in given tracking points (see Figure 3.5).

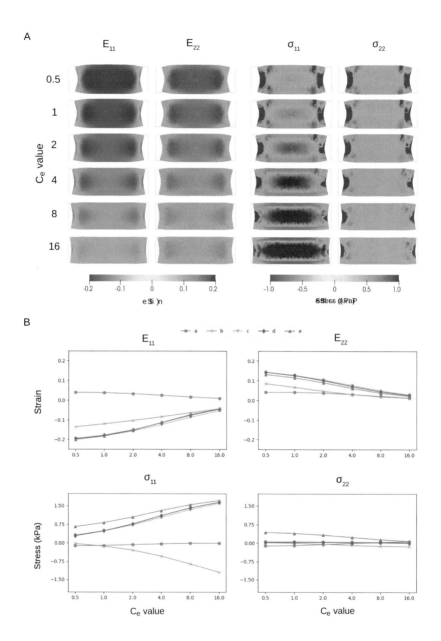

Fig. 3.9: First and middle components of the Cauchy stress tensor σ and Green-Lagrange strain E, for a single cell, as we vary the parameter C_e, which defines the stiffness of the material in the extracellular domain. Panel A shows spatial variation over a cross-section of the cell (see Figure 3.4), at peak. Panel B shows how the value, at peak, changes in given tracking points (see Figure 3.5).

the two domains. The continuity of stress across the cell membrane implies that the membrane itself has no stiffness, which is obviously incorrect, but it may be a reasonable assumption for many applications. The impact of different membrane mechanical properties should be explored further in a future study. Similarly, both the intra- and extracellular domains are assumed to be hyperelastic materials, which is probably a fairly crude approximation of the actual behaviour. In reality both of these domains are complex compositions of fluids and various embedded proteins structures, and the material behavior is most likely quite complex. Visco-elastic material models could potentially be a more accurate description than the hyper-elastic models applied here, but the required level of detail and material model complexity remains to be determined. Finally, we have here assumed that both domains are initially in a stress-free resting state, while experiments have shown that the extra-cellular matrix shrinks considerably when the myocytes are removed. Thus indicates that the resting state is actually an equilibrium state with non-zero stress in both domains, and accurately capturing the overall mechanical behaviour may require including this pre-stress in the model.

In general, the level of detail and complexity of the model formulation will be dictated by the application. Some applications may require further development of the model along the lines suggested above, while for studies of a more qualitative nature the simplest version would be sufficient. One obvious application of the developed model framework, where a fairly simple model would probably give interesting results, is to study the impact of heterogeneities in calcium concentration and mechanical properties on the contractile properties of cells and tissue.

Acknowledgements The simulations were performed on resources provided by UNINETT Sigma2 - the National Infrastructure for High Performance Computing and Data Storage in Norway. The meshes used for the simulations were generated using software provided by Miroslav Kuchta.

References

1. Ambrosi D, Pezzuto S (2012) Active stress vs. active strain in mechanobiology: constitutive issues. Journal of Elasticity 107(2):199–212
2. Campbell KS (2009) Interactions between connected half-sarcomeres produce emergent mechanical behavior in a mathematical model of muscle. PLoS computational biology 5(11):e1000560
3. Campbell SG, Lionetti FV, Campbell KS, McCulloch AD (2010) Coupling of adjacent tropomyosins enhances cross-bridge-mediated cooperative activation in a markov model of the cardiac thin filament. Biophysical journal 98(10):2254–2264
4. Chase PB, Macpherson JM, Daniel TL (2004) A spatially explicit nanomechanical model of the half-sarcomere: myofilament compliance affects ca 2+-activation. Annals of biomedical engineering 32(11):1559–1568
5. Guccione J, McCulloch A, Waldman L (1991) Passive material properties of intact ventricular myocardium determined from a cylindrical model. Journal of Biomechanical Engineering 113(1):42–55, DOI 10.1115/1.2894084
6. Holzapfel GA (2000) Nonlinear solid mechanics: a continuum approach for engineering. Wiley
7. Jæger KH, Tveito A (2020) Derivation of a cell-based mathematical model of excitable cells. In: Tveito A, Mardal KA, Rognes ME (eds) Modeling excitable tissue - The EMI framework, Simula Springer Notes in Computing, SpringerNature
8. Laadhari A, Ruiz-Baier R, Quarteroni A (2013) Fully eulerian finite element approximation of a fluid-structure interaction problem in cardiac cells. Int J Numer Meth Engng 96:712–738
9. LeGrice IJ, Smaill B, Chai L, Edgar S, Gavin J, Hunter PJ (1995) Laminar structure of the heart: ventricular myocyte arrangement and connective tissue architecture in the dog. American Journal of Physiology-Heart and Circulatory Physiology 269(2):H571–H582
10. Mijailovich SM, Nedic D, Svicevic M, Stojanovic B, Walklate J, Ujfalusi Z, Geeves MA (2017) Modeling the actin. myosin atpase cross-bridge cycle for skeletal and cardiac muscle myosin isoforms. Biophysical journal 112(5):984–996
11. Nakagome K, Sato K, Shintani SA, Ishiwata S (2016) Model simulation of the spoc wave in a bundle of striated myofibrils. Biophysics and physicobiology 13:217–226
12. Nash MP, Panfilov AV (2004) Electromechanical model of excitable tissue to study reentrant cardiac arrhythmias. Progress in Biophysics and Molecular Biology 85(2-3):501–522
13. Rice JJ, Wang F, Bers DM, de Tombe PP (2008) Approximate Model of Cooperative Activation and Crossbridge Cycling in Cardiac Muscle Using Ordinary Differential Equations. Biophysical Journal 95(5):2368–2390
14. Ruiz-Baier R, Gizzi A, Rossi S, Cherubinie C, Laadhari A, Filippi S, Quarterone A (2014) Mathematical modelling of active contraction in isolated cardiomyocytes. Mathematical medicine and biology 31:259–283

15. Sundnes J, Lines GT, Cai X, Nielsen BF, Mardal KA, Tveito A (2007) Computing the electrical activity in the heart, vol 1. Springer Science & Business Media
16. Telle Å (2020) Software for EMI – Modeling cardiac mechanics on a sub-cellular scale. DOI 10.5281/zenodo.3702168, URL https://doi.org/10.5281/zenodo.3702168
17. Tveito A, Jæger KH, Lines GT, Paszkowski Ł, Sundnes J, Edwards AG, Māki-Marttunen T, Halnes G, Einevoll GT (2017) An evaluation of the accuracy of classical models for computing the membrane potential and extracellular potential for neurons. Frontiers in computational neuroscience 11:27
18. Tveito A, Jager KH, Kuchta M, Mardal KA, Rognes ME (2017) A cell-based framework for numerical modeling of electrical conduction in cardiac tissue. Frontiers in Physics 5:48, DOI 10.3389/fphy.2017.00048, URL https://www.frontiersin.org/article/10.3389/fphy.2017.00048
19. Usyk TP, LeGrice IJ, McCulloch AD (2002) Computational model of three-dimensional cardiac electromechanics. Computing and Visualization in Science 4(4):249–257

Chapter 4

Operator Splitting and Finite Difference Schemes for Solving the EMI Model

Karoline Horgmo Jæger[1], Kristian Gregorius Hustad[1,2], Xing Cai[1,2] and Aslak Tveito[1,2]

Abstract We want to be able to perform accurate simulations of a large number of cardiac cells based on mathematical models where each individual cell is represented in the model. This implies that the computational mesh has to have a typical resolution of a few μm leading to huge computational challenges. In this paper we use a certain operator splitting of the coupled equations and show that this leads to systems that can be solved in parallel. This opens up for the possibility of simulating large numbers of coupled cardiac cells.

4.1 Introduction

In recent publications (31; 30; 13) we have shown that a cell-based model is useful for accurately representing the electrophysiology of excitable cells. Traditionally, excitable tissue is simulated based on homogenized models where the cells are not explicitly resolved, see e.g., (26; 7). In the cell-based model, we explicitly represent both the extracellular space (E), the cell membrane (M) and the intracellular space (I), and it is therefore referred to as the *EMI model*. Similar approaches to modeling excitable tissue have been used by several authors; see e.g., (2; 18; 25; 22; 24; 23; 11; 16; 34).

The EMI model is solved, numerically, using an operator splitting scheme which results in two steps; a non-linear system of ordinary differential equations (ODEs) to be solved in each computational node (i.e, degree of freedom) placed on the cell membrane, and a linear system of algebraic equations coupling the discrete Laplace equations of E and I with continuity requirements of the current over M. The spatial

---

[1] Simula Research Laboratory, Norway
[2] Department of Informatics, University of Oslo, Norway

The Author(s) 2021
A. Tveito et al. (eds.), *Modeling Excitable Tissue*, Simula SpringerBriefs on Computing 7, https://doi.org/10.1007/978-3-030-61157-6_4

resolution used in the discretization of the model is usually between 1 μm and 4 μm, thus only 1 mm^3 of tissue leads to more than 10^7 computational nodes. For an adult human cardiac cell, with a resolution of 2 μm, the number of computational nodes per cell (including the associated extracellular space) is about 6000 (see (30), Table 7). Thus, for a limited number of cells, the linear system coupling all the discrete Laplace equations is manageable. In fact, the system was solved using Matlab for up to 16,384 cells, with about 9.8×10^7 computational nodes, see (30).

However, not only the sheer size of the linear system is a challenge, also the properties of the linear system are unusual. In scientific computing, one of the most well-studied problems is solution of linear systems arising from the discretization of elliptic boundary value problems; see e.g., (5; 21; 8). Unfortunately, the EMI system does not naturally fall into the category of elliptic boundary value problems that can be solved using well-developed numerical machinery. It is therefore of importance to develop a splitting strategy for the EMI model that leads to sub-problems of the elliptic type. In (14), we showed that such a splitting can indeed be achieved. Here, we will review this convenient way of splitting the EMI model and show how to solve the system numerically using a finite difference method. Moreover, we will use the numerical scheme to assess the conduction properties in a small collection of cells where a sub-group of the cells are ischemic. Furthermore, we will present a parallel implementation of the splitting strategy, based on using open-source numerical libraries. This code is considerably faster than the existing Matlab code, and well suited for shared-memory parallel computers.

4.2 The EMI Model

We model the electrical properties of collections of cardiac cells using the EMI model introduced in (24; 1; 2; 31; 30). In Figure 4.1 we show the computational domains in the case of two coupled cells. Here, Ω_i^1 and Ω_i^2 denote the intracellular domains, and Ω_e denotes the extracellular space. The cell membranes are denoted by Γ_1 and Γ_2, respectively. The intercalated disc at the intersection between Ω_i^1 and Ω_i^2, allowing for currents between the cells, is denoted by $\Gamma_{1,2}$. With this notation at hand, the EMI model takes the following form:

$$\nabla \cdot \sigma_i \nabla u_i^k = 0 \quad \text{in } \Omega_i^k, \qquad\qquad n_e \cdot \sigma_e \nabla u_e = -n_i^k \cdot \sigma_i \nabla u_i^k \equiv I_m^k \quad \text{at } \Gamma_k,$$

$$\nabla \cdot \sigma_e \nabla u_e = 0 \quad \text{in } \Omega_e, \qquad\qquad v_t^k = \frac{1}{C_m}(I_m^k - I_{\text{ion}}^k) \quad\quad\ \text{at } \Gamma_k,$$

$$u_e = 0 \quad \text{at } \partial\Omega_e^D, \qquad\qquad u_i^k - u_i^{\tilde{k}} = w^k \quad\quad\quad\quad\ \text{at } \Gamma_{k,\tilde{k}},$$

$$n_e \cdot \sigma_e \nabla u_e = 0 \quad \text{at } \partial\Omega_e^N, \qquad\qquad n_i^{\tilde{k}} \cdot \sigma_i \nabla u_i^{\tilde{k}} = -n_i^k \cdot \sigma_i \nabla u_i^k \equiv I_{k,\tilde{k}} \ \text{at } \Gamma_{k,\tilde{k}},$$

$$u_i^k - u_e = v^k \quad \text{at } \Gamma_k, \qquad\qquad w_t^k = \frac{1}{C_g}(I_{k,\tilde{k}} - I_{\text{gap}}^k) \quad\quad \text{at } \Gamma_{k,\tilde{k}},$$

$$s_t^k = F^k \quad \text{at } \Gamma_k.$$

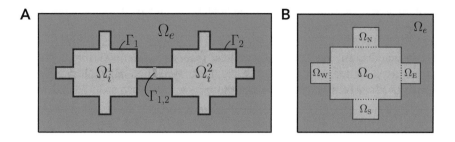

Fig. 4.1: **A**: Two-dimensional version of the EMI model domain in the case of two connected cells. Here, the cells Ω_i^1 and Ω_i^2, with cell membranes denoted by Γ_1 and Γ_2, respectively, are connected to each other by the intercalated disc, $\Gamma_{1,2}$, and surrounded by an extracellular space, denoted by Ω_e. **B**: Two-dimensional illustration of the geometry used for a single cell. The intracellular domain of each cell is composed of five subdomains Ω_O, Ω_W, Ω_E, Ω_S, and Ω_N. The sizes of the subdomains are specified in Table 4.1.

The model is stated for cell number k, and $\tilde{k}$ denotes one of the six neighboring cells (in 3D: north, west, south, east, above, below). In the model, u_e, u_i^k, and $v^k = u_i^k - u_e$ denote the extracellular, intracellular, and transmembrane potentials, respectively. Also, w^k is the potential difference across the intercalated disc[1], $\Gamma_{k,\tilde{k}}$, and σ_i and σ_e denote intracellular and extracellular conductivities, whereas C_m and C_g represent the specific capacitance of the membrane and the intercalated disc, respectively. Furthermore, n_e, n_i^k, and $n_i^{\tilde{k}}$ represent the outward pointing unit normal vectors of Ω_e, Ω_i^k and $\Omega_i^{\tilde{k}}$, respectively. A homogeneous Dirichlet boundary condition is applied at the outer extracellular boundary in the x-direction ($\partial\Omega_e^D$), and a homogeneous Neumann boundary condition is applied at the boundary in the y- and z-directions ($\partial\Omega_e^N$). The parameters used in the computations below are summarized in Table 4.1. The properties of the cell membrane and the gap junctions are represented by F, I_{ion} and I_{gap}. In the computations reported below, we use the Grandi et al. model(9), to model the dynamics of the membrane (F and I_{ion}), and for the gap junctions we use the simple passive model $I_{\text{gap}}^k = w^k / R_g$.

4.2.1 Operator Splitting Applied to the EMI Model

As mentioned above, a key step in solving the EMI model is to split the equations into parts that can be solved using standard tools. In (14), we derived a splitting scheme that leads to two key numerical challenges: Non-linear systems of ODEs to be solved

---

[1] Note that w^k is defined specifically for each cell.

Parameter	Value	Parameter	Value
Size Ω_O	$100\ \mu m \times 18\ \mu m \times 18\ \mu m$	C_m	$1\ \mu F/cm^2$
Size Ω_W, Ω_E	$4\ \mu m \times 10\ \mu m \times 10\ \mu m$	C_g	$0.5\ \mu F/cm^2$
Size Ω_N, Ω_S	$10\ \mu m \times 4\ \mu m \times 10\ \mu m$	σ_i	$4\ mS/cm$
$\Delta x, \Delta y, \Delta z$	$2\ \mu m$	σ_e	$20\ mS/cm$
Δt	$0.02\ ms$	R_g	$0.0045\ k\Omega cm^2$
Δt_{ODE}	$0.001\ ms$	M_{it}, N_{it}	2

Table 4.1: Parameter values used in the simulations, based on (13). For parameters of the Grandi model, see (9).

Algorithm 1: Summary of the splitting algorithm for the EMI model for connected cells.

Initial conditions: $v^{k,0}, s^{k,0}, w^{k,0}, u_e^0$ for all k.

for $n = 1, \ldots, N_t$:

    Step 1: For all k, find $s^{k,n}$ and $\bar{v}^k$ at the nodes of the membrane Γ_k of cell k by solving a time step Δt from $(s^{k,n-1}, v^{k,n-1})$ of

$$v_t^k = -\frac{1}{C_m} I_{ion}(v^k, s^k),$$
$$s_t^k = F(v^k, s^k).$$

    Define $\bar{u}_e = u_e^{n-1}, \bar{w}^k = w^{k,n-1}$.

    for $j = 1, \ldots, N_{it}$:

        Step 2:

        for $m = 1, \ldots, M_{it}$:

            For every k, find $\bar{u}_i^k$ by solving

$$\nabla \cdot \sigma_i \nabla \bar{u}_i^k = 0 \qquad\qquad\qquad \text{in } \Omega_i^k,$$
$$\bar{u}_i^k + \frac{\Delta t}{C_m} n_i^k \cdot \sigma_i \nabla \bar{u}_i^k = \bar{v}^k + \bar{u}_e \qquad\qquad \text{at } \Gamma_k,$$
$$-n_i^k \cdot \sigma_i \nabla \bar{u}_i^k = \frac{1}{R_g} \bar{w}^k + C_g \frac{\bar{w}^k - w^{k,n-1}}{\Delta t} \qquad \text{at } \Gamma_{k,\tilde{k}},$$

            where $\tilde{k}$ denotes each of the neighboring cells of cell k.

            Update $\bar{w}^k = \bar{u}_i^k - \bar{u}_i^{\tilde{k}}$ at $\Gamma_{k,\tilde{k}}$ for all k and $\tilde{k}$.

        end

        Step 3: Find $\bar{u}_e$ by solving

$$\nabla \cdot \sigma_e \nabla \bar{u}_e = 0 \qquad\qquad \text{in } \Omega_e,$$
$$\bar{u}_e = 0 \qquad\qquad \text{at } \partial\Omega_e^D,$$
$$n_e \cdot \sigma_e \nabla \bar{u}_e = 0 \qquad\qquad \text{at } \partial\Omega_e^N,$$
$$n_e \cdot \sigma_e \nabla \bar{u}_e = -n_i^k \cdot \sigma_i \nabla \bar{u}_i^k \qquad \text{at } \Gamma_k \text{ for all } k.$$

    end

    Define $u_e^n = \bar{u}_e, u_i^{k,n} = \bar{u}_i^k, w^k = \bar{w}^k$ for all k.

    Step 4: Define $v^{k,n} = u_i^{k,n} - u_e^n$ at Γ_k for all k.

end

---

at each computational node located at the cell membranes, and a series of elliptic equations; see Algorithm 1. All the differential equations involved in Algorithm 1 are of classical type and can be solved using well-established numerical methods. In our present implementation, we apply a straightforward finite difference scheme (see e.g., (29) for an elementary introduction to finite differences) for the elliptic equations and the Forward Euler method with a substepping time step Δt_{ODE} for solving the non-linear ODEs modeling the membrane dynamics (see (30)). However, it is worth observing that elliptic equations can as well be solved using the finite element method, or a finite volume method, thus allowing for more flexible and adaptive meshes.

4.3 Simulating the Effect of a Region of Ischemic Cells

In order to demonstrate an application of Algorithm 1 above, we consider a collection of cells where a fraction of the cells are ischemic. This is known to perturb the electrical conduction and may lead to arrhythmias; see e.g., (33; 28; 19; 27). This problem has been carefully studied using homogenized models (mostly the monodomain model), but here we will show that the ischemic regions also have local effects when only very few cells are considered. In Figure 4.2, we consider a collection of cells organized in a two-dimensional mesh of 22×12 cells. The cells are modeled using the Grandi model with parameters as stated in (9). Within the domain, 8×6 of the center cells are ischemic in the sense that the extracellular potassium concentration surrounding these cells is increased from 5.4 mM to 10 mM. For the ischemic cells, we use the steady-state values of the state variables for the increased extracellular potassium concentration as initial conditions, and for the remaining cells, we use the steady-state values of the default Grandi model. In addition, we run the simulation for 5 ms before stimulation.

In the simulation results we observe that the ischemic region slows down conductions and thus perturbs the wave in the intracellular potential moving from left to right. This is consistent with the result obtained in (6) (cf. Figure 5), where the monodomain model was used. Such perturbations are known to be arrhythmogenic and have been observed several times in numerical experiments; see e.g., (33; 19; 15; 3; 6). Here, we observe that such perturbations can be initiated locally when only a few cells are subject to surroundings with elevated potassium concentration.

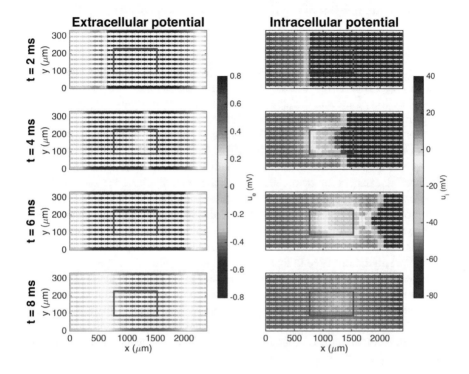

Fig. 4.2: Extracellular potential (left) and intracellular potential (right) at four differ-ent points in time in an EMI model simulation with an ischemic region in the center of the domain, marked by the purple rectangle. The parameter values used in the simulation are given in Table 4.1.

4.4 A Scalable Implementation of the Splitting Scheme

In expectation of future simulations of excitable issues that may involve a huge number of cells, we see the need of a scalable implementation of the new splitting scheme, so that it can run efficiently on parallel computers. One specific criterion is that the computation time should grow linearly with the number of cells involved. Additionally, the design goals of this new code should also include independence of proprietary software (such as Matlab) and plug-and-play of the different numerical components. This section will present a preliminary version of such a scalable implementation.

4.4.1 The Linear System for the Intracellular Potential

The main benefit of the new the splitting scheme is that the intracellular Laplace equations (one per cell) are decoupled from the extracellular Laplace equation, as stated in Algorithm 1. If we assume a constant intracellular conductivity σ_i and that each cell is of the same shape and size, as shown in Figure 4.1, the matrices arising from a standard finite difference discretization of the intracellular Laplace equations for the individual cells will be mostly identical. There are only a small number of unique intracellular matrices, depending on whether there is a neighboring cell connected to each of the intercalated discs. It is thus unnecessary to compute an intracellular matrix for each cell. Instead, the cells that have the same neighbor connectivity situation can share the same intracellular matrix. This not only reduces the memory usage of an implementation, but also improves data reuse in the caches of a computer. Moreover, since the number of computational nodes per intracellular domain is relatively small (each intracellular domain has about 5300 degrees of freedom for the simulations used in this chapter), it is very efficient to use a direct solver each time an intracellular Laplace problem needs to be solved. Specifically, the LU factorization of each unique intracellular matrix A_I can be pre-calculated, which renders the solution of $A_I \bar{u}_i^k = b_i^k$ per cell to be merely invoking the forward-backward substitution procedure. Parallelism of the computation mainly arises from the fact that the intracellular Laplace equations can be solved independently of each other, while limited parallelism also exists within each forward-backward substitution.

4.4.2 The Linear System for the Extracellular Potential

For the overall extracellular Laplace problem, which can be huge depending on the spatial resolution and the total number of cells, an iterative solver is more appropriate. Take for instance the case of 128×128 cells. The corresponding discrete extracellular Laplace equation has $107,202,214$ degrees of freedom. Independent of the spatial resolution and the number of cells involved, the extracellular matrix A_E arising from a standard finite difference discretization is symmetric and positive-definite (some care is needed to discretize the boundary conditions on the membranes). The resulting linear system $A_E \bar{u}_e = b_e$ is thus a perfect candidate for the conjugate gradients (CG) method with an algebraic multigrid (AMG) preconditioner. Under optimal conditions, an AMG preconditioner requires a constant number of iterations to reach convergence independent of the linear system size, although the number of grid levels inside the AMG preconditioner may increase with the system size. Parallelism readily exists in iterative solvers, with several software libraries providing parallel implementations of CG and AMG.

4.4.3 The Non-Linear ODE System for the Membrane Potential

For solving the non-linear ODE system per computational node on the membranes, a straightforward and often very efficient numerical strategy is the Forward Euler method with a substepping time step Δt_{ODE}. Since the non-linear ODE system on each membrane node is independent of the others, the ODE computation possesses the most ample parallelism.

4.4.4 The Implementation

The Python programming language has been chosen for the implementation, mostly because of its flexibility for interfacing with numerical software libraries written in performance-friendly languages such as C and C++. We have used the `ctypes` module from the standard Python library for this purpose. The choice of Python also simplified a partial translation from the existing MATLAB code developed in (14).

We have chosen the SuperLU library (17) for performing the LU factorization of the intracellular matrices and the subsequent forward-backward substitution, via the bindings that are provided by SciPy (32). For the extracellular Laplace equation, we have used the ViennaCL library (20) for its implementation of CG and AMG. The CG iterations are by default configured to terminate when a tolerance of 10^{-5} is reached. The AMG preconditioner has been configured to use the maximum independent set (MIS), see (4), as the coarsening algorithm and smoothed aggregation as the interpolation algorithm. For the ODE part, the Gotran automated code generator (10) has been used to translate the Grandi cell model into C code, callable from the Python side.

4.4.5 Parallelization

The Python implementation currently relies on the adopted numerical libraries (SuperLU and ViennaCL) for an implicit parallelization of the PDE computation through multi-threading. This form of parallelization suits for shared-memory parallel computers, such as laptops or servers that use multicore CPUs. Multi-threading of the ODE computation is also enabled by inserting OpenMP compiler directives into the C code that is generated automatically by Gotran. The advantage of this implicit parallelization is that the user does not have to care about parallelization-specific coding. The downside is that all the computations have to run on a shared-memory system. It is possible to achieve the more general parallelization that targets distributed-memory parallel computers, which will be a task for future work.

4.4.6 Performance Results

The simulations in this section were run on a dual-socket server with two 32-core AMD EPYC 7601 CPUs, each with 8-channel memory operating at 2666 MT/s. The number of OpenMP threads was set by default to the number of logical cores, equaling 128. Moreover, the environment variable `OMP_PROC_BIND=TRUE` was set to prevent the threads from migrating between the cores (which typically leads to unnecessary performance loss).

Table 4.2 shows the average solution time per time step for the 10 first time steps, where all the parameters are as prescribed in Table 4.1. The number of cardiac cells is doubled in the x and y directions for each row, and we observe that the time per cell remains fairly constant, indicating that the time to solution is a linear function of the number of cardiac cells simulated.

Cells	time usage for all cells (s)				time per cell (ms)			
	E	M	I	total	E	M	I	total
4×4	0.38	0.03	0.09	0.50	24.0	2.2	5.3	31.5
8×8	1.54	0.11	0.34	1.99	24.1	1.8	5.2	31.2
16×16	2.27	0.45	1.20	3.92	8.8	1.7	4.7	15.3
32×32	8.91	1.72	4.98	15.61	8.7	1.7	4.9	15.2
64×64	30.46	6.73	19.15	56.33	7.4	1.6	4.7	13.8
128×128	123.73	30.60	72.83	227.16	7.6	1.9	4.4	13.9

Table 4.2: Average solution time per time step for the E, M and I domains.

4.5 Software

The Matlab code used to compute the solutions shown in Figure 4.2 and the Python code discussed in Section 4.4 can be found at `https://github.com/KGHustad/emi-book-2020-splitting-code`. An archived version (12) is also available.

4.6 Conclusion

In this chapter we have presented a numerical scheme for solving the EMI equations using operator splitting. The scheme allows for parallel solution of individual cells combined with a global solution of the equation modeling the extracellular potential.

The latter is well suited for using optimal linear solvers such as AMG. The overall code scales linearly with the number cells and thus allows for simulation of a large number of cells. It remains to be seen how well this will work for very large numbers of cells; this is subject for further work.

Acknowledgements The research presented in this chapter has benefited from the Experimental Infrastructure for Exploration of Exascale Computing (eX3), which is financially supported by the Research Council of Norway under contract 270053.

References

1. Agudelo-Toro A (2012) Numerical simulations on the biophysical foundations of the neuronal extracellular space. PhD thesis, Niedersächsische Staats-und Universitätsbibliothek Göttingen
2. Agudelo-Toro A, Neef A (2013) Computationally efficient simulation of electrical activity at cell membranes interacting with self-generated and externally imposed electric fields. Journal of Neural Engineering 10(2):026019
3. Alonso S, Bär M, Echebarria B (2016) Nonlinear physics of electrical wave propagation in the heart: a review. Reports on Progress in Physics 79(9):096601
4. Bell N, Dalton S, Olson LN (2012) Exposing fine-grained parallelism in algebraic multigrid methods. SIAM Journal on Scientific Computing 34(4):C123–C152, DOI 10.1137/110838844
5. Benzi M (2002) Preconditioning techniques for large linear systems: a survey. Journal of computational Physics 182(2):418–477
6. Dutta S, Mincholé A, Quinn TA, Rodriguez B (2017) Electrophysiological properties of computational human ventricular cell action potential models under acute ischemic conditions. Progress in biophysics and molecular biology 129:40–52
7. Franzone PC, Pavarino LF, Scacchi S (2014) Mathematical Cardiac Electrophysiology. Springer International Publishing
8. Gergelits T, Mardal KA, Nielsen BF, Strakos Z (2019) Laplacian preconditioning of elliptic pdes: Localization of the eigenvalues of the discretized operator. SIAM Journal on Numerical Analysis 57(3):1369–1394
9. Grandi E, Pasqualini FS, Bers DM (2010) A novel computational model of the human ventricular action potential and Ca transient. Journal of Molecular and Cellular Cardiology 48:112–121
10. Hake J, Finsberg H, Hustad KG, Bahij G (2020) Gotran – General ODE TRANslator. https://github.com/ComputationalPhysiology/gotran
11. Hogues H, Leon LJ, Roberge FA (1992) A model study of electric field interactions between cardiac myocytes. IEEE Transactions on Biomedical Engineering 39(12):1232–1243
12. Jæger KH, Hustad KG (2020) Supplementary material (code) for the chapter "Operator splitting and finite difference schemes for solving the EMI model" appearing in "EMI: Cell-based Mathematical Model of Excitable Cells". DOI 10.5281/zenodo.3707472, URL https://doi.org/10.5281/zenodo.3707472
13. Jæger KH, Edwards AG, McCulloch A, Tveito A (2019) Properties of cardiac conduction in a cell-based computational model. PLoS computational biology 15(5):e1007042
14. Jæger KH, Hustad KG, Cai X, Tveito A (2020) Efficient numerical solution of the EMI model representing the extracellular space (E), cell membrane (M) and intracellular space (I) of a collection of cardiac cells. Preprint
15. Kazbanov IV, Clayton RH, Nash MP, Bradley CP, Paterson DJ, Hayward MP, Taggart P, Panfilov AV (2014) Effect of global cardiac ischemia on human ventricular fibrillation: in-

sights from a multi-scale mechanistic model of the human heart. PLoS computational biology 10(11):e1003891

16. Krassowska W, Neu JC (1994) Response of a single cell to an external electric field. Biophysical Journal 66(6):1768–1776

17. Li XS (2005) An overview of SuperLU: Algorithms, implementation, and user interface. ACM Trans Math Softw 31(3):302–325

18. Roberts SF, Stinstra JG, Henriquez CS (2008) Effect of nonuniform interstitial space properties on impulse propagation: a discrete multidomain model. Biophysical Journal 95(8):3724–3737

19. Romero L, Trénor B, Alonso JM, Tobón C, Saiz J, Ferrero JM (2009) The relative role of refractoriness and source–sink relationship in reentry generation during simulated acute ischemia. Annals of Biomedical Engineering 37(8):1560–1571

20. Rupp K, Tillet P, Rudolf F, Weinbub J, Morhammer A, Grasser T, JÃŒEngel A, Selberherr S (2016) ViennaCL—linear algebra library for multi- and many-core architectures. SIAM Journal on Scientific Computing 38(5):S412–S439, DOI 10.1137/15M1026419

21. Smith B, Bjorstad P, Gropp W (2004) Domain decomposition: parallel multilevel methods for elliptic partial differential equations. Cambridge university press

22. Stinstra J, MacLeod R, Henriquez C (2010) Incorporating histology into a 3D microscopic computer model of myocardium to study propagation at a cellular level. Annals of Biomedical Engineering 38(4):1399–1414

23. Stinstra JG, Hopenfeld B, MacLeod RS (2005) On the passive cardiac conductivity. Annals of Biomedical Engineering 33(12):1743–1751

24. Stinstra JG, Roberts SF, Pormann JB, MacLeod RS, Henriquez CS (2006) A model of 3D propagation in discrete cardiac tissue. In: Computers in Cardiology, 2006, IEEE, pp 41–44

25. Stinstra JG, Henriquez CS, MacLeod RS (2009) Comparison of microscopic and bidomain models of anisotropic conduction. In: Computers in Cardiology, IEEE, pp 657–660

26. Sundnes J, Lines G, Cai X, Nielsen B, Mardal KA, Tveito A (2006) Computing the Electrical Activity of the Heart. Springer

27. Tveito A, Lines G (2008) A condition for setting off ectopic waves in computational models of excitable cells. Math Biosci 213:141–150

28. Tveito A, Lines GT (2009) A note on a method for determining advantageous properties of an anti-arrhythmic drug based on a mathematical model of cardiac cells. Mathematical Biosciences 217(2):167–173, DOI DOI:10.1016/j.mbs.2008.12.001, URL http://www.sciencedirect.com/science/article/B6VHX-4V70RG1-1/2/8b8ae8d1fbf9e2c74235b7e7a97c6f6e

29. Tveito A, Langtangen HP, Nielsen BF, Cai X (2010) Elements of scientific computing, vol 7. Springer Science & Business Media

30. Tveito A, Jæger KH, Kuchta M, Mardal KA, Rognes ME (2017) A cell-based framework for numerical modeling of electrical conduction in cardiac tissue. Frontiers in Physics 5:48

31. Tveito A, Jæger KH, Lines GT, Paszkowski Ł, Sundnes J, Edwards AG, Mäki-Marttunen T, Halnes G, Einevoll GT (2017) An evaluation of the accuracy of classical models for computing the membrane potential and extracellular potential for neurons. Frontiers in Computational Neuroscience 11:27

32. Virtanen P, Gommers R, Oliphant TE, Haberland M, Reddy T, Cournapeau D, Burovski E, Peterson P, Weckesser W, Bright J, van der Walt SJ, Brett M, Wilson J, Jarrod Millman K, Mayorov N, Nelson ARJ, Jones E, Kern R, Larson E, Carey C, Polat i, Feng Y, Moore EW, Vand erPlas J, Laxalde D, Perktold J, Cimrman R, Henriksen I, Quintero EA, Harris CR, Archibald AM, Ribeiro AH, Pedregosa F, van Mulbregt P, Contributors S (2020) SciPy 1.0: Fundamental Algorithms for Scientific Computing in Python. Nature Methods 17:261–272, DOI https://doi.org/10.1038/s41592-019-0686-2

33. Xie F, Qu Z, Garfinkel A, Weiss JN (2001) Effects of simulated ischemia on spiral wave stability. American Journal of Physiology-Heart and Circulatory Physiology 280(4):H1667–H1673

34. Ying W, Henriquez CS (2007) Hybrid finite element method for describing the electrical response of biological cells to applied fields. IEEE Transactions on Biomedical Engineering 54(4):611–620

Chapter 5

Solving the EMI Equations using Finite Element Methods

Miroslav Kuchta[1], Kent-André Mardal[1,2] and Marie E. Rognes[1]

Abstract This chapter discusses 2×2 symmetric variational formulations and associated finite element methods for the EMI equations. We demonstrate that the presented methods converge at expected rates, and compare the approaches in terms of approximation of the transmembrane potential. Overall, the choice of which formulation to employ for solving EMI models becomes largely a matter of desired accuracy and available computational resources.

5.1 Introduction

In this chapter, we present different weak formulations and corresponding finite element methods for solving the EMI equations as presented in (7, Chapter 1) over a physiological cell Ω_i and its membrane Γ surrounded by an extracellular space Ω_e and a time interval $(0, T]$ for some time $T > 0$. This coupled, time-dependent, and typically nonlinear system of equations can be targeted numerically by an operator splitting scheme, see e.g (8, Chapter 4). Such an approach, combined with for instance an implicit Euler discretization in time, gives the following stationary and linear, but still coupled system of equations to be solved at each time-step: find the potentials $u_e = u_e(x)$, $u_i = u_i(x)$ (and current $I_m = I_m(x)$) such that

[1]Simula Research Laboratory, Norway
[2]Department of Mathematics, University of Oslo, Oslo, Norway

The Author(s) 2021
A. Tveito et al. (eds.), *Modeling Excitable Tissue*, Simula SpringerBriefs
on Computing 7, https://doi.org/10.1007/978-3-030-61157-6_5

$$-\nabla \cdot (\sigma_e \nabla u_e) = 0 \qquad\qquad \text{in } \Omega_e, \qquad (5.1a)$$

$$-\nabla \cdot (\sigma_i \nabla u_i) = 0 \qquad\qquad \text{in } \Omega_i \qquad (5.1b)$$

$$\sigma_e \nabla u_e \cdot \boldsymbol{n}_e = -\sigma_i \nabla u_i \cdot \boldsymbol{n}_i \equiv I_m \qquad \text{on } \Gamma, \qquad (5.1c)$$

$$u_i - u_e = v \qquad\qquad \text{on } \Gamma, \qquad (5.1d)$$

$$v - C_m^{-1} \Delta t I_m = f \qquad\qquad \text{on } \Gamma, \qquad (5.1e)$$

where $\Delta t > 0$ denotes a time step size, and $\boldsymbol{n}_e$ (resp. $\boldsymbol{n}_i$) denotes the outward pointing normal on Γ when viewed from Ω_e (resp. Ω_i). In our (implicit Euler) time discretization context, the known right-hand side f of (5.1e) combines the previous transmembrane potential solution, v_0, and the evaluation of the ionic current, I_{ion}, into $f \equiv v_0 - C_m^{-1} \Delta t I_{\text{ion}}$.

We assume that the potential is grounded on part of the external boundary Γ_e^D and that the remaining external boundary Γ_e^N is insulated. These assumptions give the boundary conditions:

$$u_e = 0 \qquad\qquad \text{on } \Gamma_e^D, \qquad (5.2a)$$

$$\sigma_e \nabla u_e \cdot \boldsymbol{n}_e = 0 \qquad\qquad \text{on } \Gamma_e^N. \qquad (5.2b)$$

This geometrical setting is illustrated in Figure 5.1.

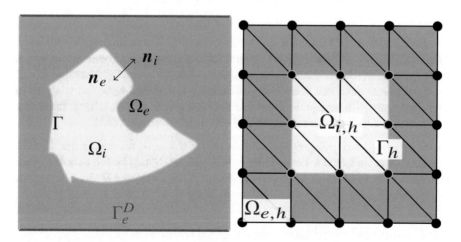

Fig. 5.1: (Left) Illustration of the geometric setting for the single cell EMI problem. (Right) Sample meshes for our benchmark problem (5.17). The boundary facets of the intracellular mesh $\Omega_{i,h}$ form the membrane mesh Γ_h.

Remark 5.1 We remark that a single cell model is here considered for simplicity. Indeed the formulations to be studied below can be similarly derived for the intercalated model (collections of connected cells). Formulations for a number of disconnected cells are then practically identical to the case considered here.

Remark 5.2 If (5.2) is considered without any Dirichlet boundary data, i.e. $|\Gamma_e^D| = 0$, then only the transmembrane potential is fixed and the intracellular and extracellular potentials are determined only up to a single, common constant.

The EMI equations (5.1) set a rich scene for numerical exploration and can be solved in a multitude of ways. In this chapter, we will derive 2×2 different weak formulations (each defining a finite element method) of this system. The two first formulations (in Section 5.2) compute the intracellular and extracellular potentials as the main unknowns. These are referred to as *primal formulations*. The latter two formulations (in Section 5.3) additionally introduce the current densities $J_i = -\sigma_i \nabla u_i$ and $J_e = -\sigma_e \nabla u_e$ as independent unknowns. These are referred to as *mixed formulations*. We compare finite element discretizations of the primal and mixed formulations with respect to the approximation of the transmembrane potential v in Section 5.4. This choice is motivated by the observation that v is closely coupled to the membrane dynamics as discussed in Chapter 1.

5.1.1 Preliminaries: Function Spaces and Norms

The EMI equations (5.1) define a multi-dimensional[1] PDE system coupling unknown fields defined over cellular domains and fields defined over the cell membrane, which can be viewed as a lower-dimensional manifold. Identifying the right function spaces for the different unknown fields is key to defining well-posed weak formulations of these equations. We here present suitable Sobolev spaces for this setting; the reader is referred to e.g. (3; 5) for more material and careful formalizations.

Let Ω be a bounded, polygonal domain in $\mathbb{R}^d$ for $d = 2, 3$. We denote the space of square-integrable functions over Ω by $L^2(\Omega)$, and let $H^1(\Omega)$ be the Sobolev space of functions in $L^2(\Omega)$ with weak derivatives in $L^2(\Omega)$. The space $\boldsymbol{H}(\text{div}, \Omega)$ contains vector-valued functions $\boldsymbol{v} : \Omega \to \mathbb{R}^d$ such that $\boldsymbol{v} \in \boldsymbol{L}^2(\Omega)$ and $\nabla \cdot \boldsymbol{v} \in L^2(\Omega)$. In general, when clear from the context, the domain will be omitted from the notation.

The L^2-inner product and norm for $u, v \in L^2(\Omega)$ is written as

$$(u, v)_{0,\Omega} = \int_\Omega uv \, dx, \quad \|v\|_{0,\Omega}^2 = \int_\Omega v^2 \, dx.$$

Similarly, we define the H^1-norm as $\|v\|_{1,\Omega}^2 = \|v\|_{0,\Omega}^2 + \|\nabla v\|_{0,\Omega}^2$ for $v \in H^1(\Omega)$, and the $\boldsymbol{H}(\text{div})$-norm as $\|\boldsymbol{v}\|_{\text{div},\Omega}^2 = \|\boldsymbol{v}\|_{0,\Omega}^2 + \|\nabla \cdot \boldsymbol{v}\|_{0,\Omega}^2$.

---

[1] PDEs coupling fields over domains of different topological dimensions are often referred to as mixed-dimensional PDEs. To avoid the confusion-inducing term mixed-dimensional mixed in the subsequent sections, we instead use the term multi-dimensional in this chapter.

For $\Gamma \subseteq \partial\Omega$, we define the constrained spaces $H^1_\Gamma(\Omega) = \{v \in H^1(\Omega) \mid v = 0 \text{ on } \Gamma\}$, and $H_\Gamma(\text{div}, \Omega) = \{v \in H(\text{div}, \Omega) \mid v \cdot n = 0 \text{ on } \Gamma\}$ where n is the (outward pointing) normal vector of Γ. Finally the spaces $H^{1/2}(\Gamma)$ and $H^{-1/2}(\Gamma)$ are the trace spaces of H^1 and $H(\text{div})$ respectively (6, Ch. 1., 2.). Here, the spaces will be considered with the norm defined in terms of fractional powers of the Helmholtz operator, see e.g. (4), i.e.

$$\|u\|^2_s = (u, (-\Delta + I)^s u)_{0,\Gamma}, \, u \in C^\infty(\Gamma).$$

We remark that in the following experiments the fractional norm is evaluated using the eigenvalue decomposition of $-\Delta + I$ as detailed in (11).

5.2 Primal Formulations

We present two primal formulations of the stationary EMI system (5.1) with the boundary conditions given by (5.2): one *single-dimensional* formulation and one *multi-dimensional* formulation. The difference in the intra- and extracellular potential across the cell membrane Γ sets up a potential jump, the transmembrane potential v, c.f. (5.1d). Due to this jump, one cannot define a global, differentiable potential $u \in H^1(\Omega_i \cup \Omega_e)$ such that $u|_{\Omega_i} = u_i$ and $u|_{\Omega_e} = u_e$. Instead, we seek $u_i \in H^1(\Omega_i)$ and $u_e \in H^1(\Omega_e)$ separately. In the single-dimensional formulation, these are the only unknown fields, while in the multi-dimensional formulation, we keep I_m as an additional unknown.

5.2.1 Single-Dimensional Primal Formulation

Define the function spaces

$$V_i = H^1(\Omega_i), \quad V_e = H^1_{\Gamma^D_e}(\Omega_e). \tag{5.3}$$

To derive a weak formulation of (5.1), multiply (5.1a) by a test function $v_e \in V_e$, (5.1b) by another test function $v_i \in V_i$, and integrate the divergence by parts. This yields the variational formulation: find $u_e \in V_e$, $u_i \in V_i$ satisfying

$$\int_{\Omega_e} \sigma_e \nabla u_e \cdot \nabla v_e \, dx - \int_\Gamma \sigma_e \nabla u_e \cdot n_e v_e \, ds = 0, \tag{5.4a}$$

$$\int_{\Omega_i} \sigma_i \nabla u_i \cdot \nabla v_i \, dx + \int_\Gamma (-\sigma_i \nabla u_i \cdot n_i) v_i \, ds = 0. \tag{5.4b}$$

for all $v_e \in V_e$, $v_i \in V_i$. In the bracketed term of (5.4b), we recognize the membrane current density I_m as defined by (5.1c), and similarly, the interface contribution in the corresponding extracellular equation (5.4a) hides $-I_m$. Combining (5.1e) and (5.1d), we find that

$$I_m = C_m(\Delta t)^{-1}((u_i - u_e) - f).$$ (5.5)

After substituting (5.5) into (5.4), the single-dimensional primal weak form of (5.1) reads: find $u_i \in V_i$ and $u_e \in V_e$ such that

$$\int_{\Omega_e} \sigma_e \nabla u_e \cdot \nabla v_e \, dx + \int_\Gamma C_m(\Delta t)^{-1} u_e v_e \, ds - \int_\Gamma C_m(\Delta t)^{-1} u_i v_e \, ds =$$
$$- C_m(\Delta t)^{-1} \int_\Gamma f v_e \, ds,$$
$$\int_{\Omega_i} \sigma_i \nabla u_i \cdot \nabla v_i \, dx + \int_\Gamma C_m(\Delta t)^{-1} u_i v_i \, ds - \int_\Gamma C_m(\Delta t)^{-1} u_e v_i \, ds =$$
$$C_m(\Delta t)^{-1} \int_\Gamma f v_i \, ds,$$ (5.6)

for all $v_e \in V_e$ and $v_i \in V_i$.

We remark that (5.6) can be viewed as a coupling of two Poisson problems with a Robin boundary condition on Γ. The well-posedness of the problem is then discussed in (10, Chapter 6). Finally, the transmembrane potential can here be computed from its definition (5.1d) as a difference of the computed potentials.

5.2.2 Multi-Dimensional Primal Formulation

An alternative formulation can be derived by keeping I_m as a separate unknown field. Since Γ is of a different (lower) dimension than Ω_i, Ω_e; and as $I_m : \Gamma \to \mathbb{R}$ while $u_i : \Omega_i \to \mathbb{R}$, $u_e : \Omega_e \to \mathbb{R}$, we will refer to this as a multi-dimensional primal formulation. Observe that (5.4) now yields two equations for three unknowns $u_i \in V_i$, $u_e \in V_e$, and $I_m \in Q$:

$$\int_{\Omega_e} \sigma_e \nabla u_e \cdot \nabla v_e \, dx - \int_\Gamma I_m v_e \, ds = 0, \quad \forall v_e \in V_e,$$
$$\int_{\Omega_i} \sigma_i \nabla u_i \cdot \nabla v_i \, dx + \int_\Gamma I_m v_i \, ds = 0, \quad \forall v_i \in V_i.$$

However, the missing equation can be obtained from (5.5). Let

$$Q = H^{1/2}(\Gamma), \quad Q^* = H^{-1/2}(\Gamma).$$ (5.7)

We remind the reader that if Γ is a co-dimensional 1 subset of Ω then trace operations from Ω to Γ, $Tu = u|_\Gamma$, $u \in C(\Omega)$ and $T_n \tau = \tau|_\Gamma \cdot n$, $\tau \in C(\Omega)$, formally have the following mapping properties $T : H^1(\Omega) \to H^{1/2}(\Gamma)$ and $T_n : H(\text{div}, \Omega) \to$

$H^{-1/2}(\Gamma)$. Hence, let j_m be a a test function from Q^*. We shall then enforce that (5.5) holds in the weak sense:

$$\int_\Gamma (u_i - u_e) j_m \, ds - \int_\Gamma \Delta t C_m^{-1} I_m j_m \, ds = \int_\Gamma f j_m \, ds, \quad \forall j_m \in Q^*.$$

In turn, the multi-dimensional primal formulation of (5.1) reads: find $u_i \in V_i$, $u_e \in V_e$, $I_m \in Q^*$ such that

$$\int_{\Omega_e} \sigma_e \nabla u_e \cdot \nabla v_e \, dx - \int_\Gamma v_e I_m \, ds = 0,$$

$$\int_{\Omega_i} \sigma_i \nabla u_i \cdot \nabla v_i \, dx + \int_\Gamma v_i I_m \, ds = 0, \qquad (5.8)$$

$$\int_\Gamma -u_e j_m \, ds + \int_\Gamma u_i j_m \, ds - \int_\Gamma \Delta t C_m^{-1} I_m j_m \, ds = \int_\Gamma f j_m \, ds,$$

for all $v_i \in V_i$, $v_e \in V_e$ and $j_m \in Q^*$. We remark that (5.8) is closely related to the Babuška problem for enforcing boundary conditions by Lagrange multipliers (1) and the mortar finite element method, see e.g. (13). With regards to evaluation of the transmembrane potential, we note that v can be post-computed in several ways: from (5.1d) (as for the single-dimensional primal formulation (5.6)) or from I_m and (5.1e).

5.3 Mixed Formulations

We now turn to consider *mixed formulations* of the EMI system (5.1). Let us (re)introduce the current densities

$$J_i = -\sigma_i \nabla u_i, \quad J_e = -\sigma_e \nabla u_e \qquad (5.9)$$

and the global field J on $\Omega = \Omega_i \cup \Omega_e$ such that $J|_{\Omega_i} = J_i$ and $J|_{\Omega_e} = J_e$. In general, we use the convention that for a scalar or vector field u defined on Ω, the restriction on Ω_i and Ω_e is denoted by u_i and u_e, respectively.

With these definitions, (5.1a)–(5.1c) become: find the current densities J_i, J_e (or J) and the potentials u_i, u_e (or u) satisfying

$$\sigma_e^{-1} J_e + \nabla u_e = 0 \qquad \qquad \text{on } \Omega_e, \qquad (5.10a)$$

$$\sigma_i^{-1} J_i + \nabla u_i = 0 \qquad \qquad \text{on } \Omega_i, \qquad (5.10b)$$

$$-\nabla \cdot J = 0 \qquad \qquad \text{in } \Omega, \qquad (5.10c)$$

$$J_e \cdot n_e + J_i \cdot n_i = 0 \qquad \qquad \text{on } \Gamma. \qquad (5.10d)$$

We refer to (5.10) together with (5.1d)–(5.1e) as the mixed EMI system with boundary conditions given by (5.2). Weak formulations of the mixed form can enjoy improved conservation properties and stability properties (3). In particular, approximations of J may be computed such that they are exactly divergence free, cf. (5.10c).

The continuity condition (5.10d) ensures that the normal component of J is continuous on Γ. We remark that $v \in H(\text{div}, \Omega)$ implies continuity of $v \cdot n$ on Γ. Moreover, we observe that (5.10c) involves only divergence of the field J. It is therefore sufficient to seek J in $S = H_{\Gamma_e^N}(\text{div}, \Omega)$. Note that in contrast to the primal formulation, here the Neumann boundary condition (5.2b) is enforced as an essential condition; that is, it is included in the construction of the function space S.

5.3.1 Single-Dimensional Mixed Formulation

Let

$$S = \left\{ J \in H_{\Gamma_e^N}(\text{div}, \Omega); J \cdot n \in L^2(\Gamma) \right\}, \quad V = L^2(\Omega). \tag{5.11}$$

To derive a weak form of the mixed EMI system, consider a test function $\tau \in S$. Taking the dot product of (5.10a), (5.10b) with τ_i, τ_e, integrating and applying integration by parts then yields

$$\int_{\Omega_e} \sigma_e^{-1} J_e \cdot \tau_e \, dx - \int_{\Omega_e} u_e \nabla \cdot \tau_e \, dx + \int_\Gamma u_e \tau_e \cdot n_e \, ds = - \int_{\Gamma_e^D} u_e \tau_e \cdot n_e \, ds,$$

$$\int_{\Omega_i} \sigma_i^{-1} J_i \cdot \tau_i \, dx - \int_{\Omega_i} u_i \nabla \cdot \tau_i \, dx + \int_\Gamma u_i \tau_i \cdot n_i \, ds = 0.$$

Observe that by continuity of the normal component of the test function ($\tau_i \cdot n = \tau_e \cdot n$ on Γ), and the identity $n_e = -n_i$, the integrals on Γ can be added, resulting in $\int_\Gamma (u_i - u_e)\tau \cdot n_i$. Moreover, using (5.5), the membrane term can be rewritten as $\int_\Gamma \left(C_m^{-1} \Delta t J \cdot n_i + f \right) \tau \cdot n_i$. In turn, we arrive at the variational problem: find $J \in S$, $u \in V$ such that

$$\int_\Omega \sigma^{-1} J \cdot \tau \, dx + \int_\Gamma C_m^{-1} \Delta t J \cdot n_i \tau \cdot n_i \, ds - \int_\Omega u \nabla \cdot \tau \, dx = - \int_\Gamma f \tau \cdot n_i \, ds,$$

$$- \int_\Omega q \nabla \cdot J \, dx = 0,$$

$$\tag{5.12}$$

for all $\tau \in S$ and $q \in V$, with σ defined naturally as $\sigma|_{\Omega_i} = \sigma_i$ and likewise for Ω_e. Note that due to the extra trace regularity of the trial/test space S all the terms in (5.12), and in particular the interface term $\int_\Gamma C_m^{-1} \Delta t J \cdot n_i \tau \cdot n_i \, ds$, are well defined. Without the extra regularity, i.e. if $S = H_{\Gamma_e^N}(\text{div}, \Omega)$, this would not be the case.

We remark that (5.12) is a Γ-perturbed mixed formulation of the Poisson problem (see e.g. (3; 9) for more details on mixed formulations of the Poisson problem).

Considering the task of approximating the transmembrane potential, we observe that v can be computed in two ways, as for the multi-dimensional primal formulation. Indeed, in addition to the identity $v = u_i - u_e$, cf. (5.1d), equation (5.1e) can be used since $I_m = \mathbf{J} \cdot \mathbf{n}_i$ is readily available.

5.3.2 Multi-Dimensional Mixed Formulation

As for the primal formulations, the multi-dimensional mixed formulation is obtained by keeping the interface term as an explicit unknown field. Let

$$S = \mathbf{H}_{\Gamma_e^N}(\mathrm{div}, \Omega), \quad V = L^2(\Omega), \quad W = H^{1/2}(\Gamma). \tag{5.13}$$

To complete the formulation, the equation to be enforced weakly by test functions $w \in W$ is the membrane dynamics condition (5.1e) written in the form

$$\mathbf{J} \cdot \mathbf{n}_i - C_m(\Delta t)^{-1} v = -C_m(\Delta t)^{-1} f \quad \text{on } \Gamma.$$

The final multi-dimensional mixed weak formulation then reads: Find the current densities $\mathbf{J} \in S$, potentials $u \in V$, and transmembrane potential $v \in W$ such that

$$\int_\Omega \sigma^{-1}\mathbf{J} \cdot \boldsymbol{\tau} \, \mathrm{d}x - \int_\Omega u \nabla \cdot \boldsymbol{\tau} \, \mathrm{d}x + \int_\Gamma v\boldsymbol{\tau} \cdot \mathbf{n}_i \, \mathrm{d}s = 0,$$

$$- \int_\Omega q \nabla \cdot \mathbf{J} \, \mathrm{d}x = 0, \tag{5.14}$$

$$\int_\Gamma w\mathbf{J} \cdot \mathbf{n}_i \, \mathrm{d}s - \int_\Gamma C_m(\Delta t)^{-1} vw \, \mathrm{d}s = -C_m(\Delta t)^{-1} \int_\Gamma fw \, \mathrm{d}s,$$

for all $\boldsymbol{\tau} \in S$, $q \in V$ and $w \in W$. Note that (5.14) is defined on the standard $\mathbf{H}(\mathrm{div})$ space, cf. (5.12), as the formulation no longer contains the troublesome interface term $\int_\Gamma C_m^{-1}\Delta t \mathbf{J} \cdot \mathbf{n}_i \boldsymbol{\tau} \cdot \mathbf{n}_i \, \mathrm{d}s$. With regards to the approximation of v in formulation (5.14), observe that no post-processing is required to obtain this quantity. This is contrast to the previous three formulations. We remark that (5.14) is closely connected to the Babuška problem for the mixed Poisson equation (2).

5.4 Finite Element Spaces and Methods

To solve the primal and mixed, single- and multi-dimensional weak formulations numerically, we approximate the continuous function spaces by discrete finite element spaces. Each choice of formulation and finite element space defines a finite element method for solving the EMI system.

Now, let Ω_h be a mesh of the domain $\Omega = \Omega_i \cup \Omega_e$ with characteristic mesh size h, which conforms to Γ in the sense that no element of Ω_h has its interior intersected by Γ. The meshes $\Omega_{e,h}$ and $\Omega_{i,h}$ of the extracellular and intracellular domains are formed as non-overlapping subsets of the cells of Ω_h. As a consequence, the mesh Γ_h of Γ is formed by the facets of elements of Ω_h, cf. Figure 5.1. We remark that the single-dimensional primal formulation allows for independent discretizations of Ω_i, Ω_e as well as Γ.

The choice of the finite element spaces plays a crucial role for the stability of the different discrete formulations. In particular, for the saddle-point systems, the spaces must be compatible in the sense of Babuška-Brezzi and satisfy discrete inf-sup conditions, see e.g. (3). For the primal formulations (5.4) and (5.8), we seek discrete unknowns and test functions in

$$V_{i,h} = P_1(\Omega_{i,h}) \subset V_i, \quad V_{e,h} = P_1(\Omega_{e,h}) \subset V_e, \quad Q_h = P_1(\Gamma_h) \subset Q, \qquad (5.15)$$

where P_1 denotes the space of continuous piecewise linears (defined relative to the relevant mesh). With these spaces, we expect linear convergence with the mesh size h for all variables in H^1-norms and quadratic convergence in the L^2-norm.

For the mixed formulations (5.10) and (5.12), we seek discrete unknowns and test functions in

$$S_h = RT_0(\Omega_h) \subset S, \quad V_h = P_0(\Omega_h) \subset V, \quad W_h = P_0(\Gamma_h). \qquad (5.16)$$

Here RT_0 denotes the lowest order Raviart-Thomas finite element spaces and P_0 denotes the space of piecewise constants defined relative to the relevant mesh. These spaces satisfy the relevant stability conditions, and we expect linear convergence of all unknown fields in their respective natural norms.

5.5 Numerical Results

5.5.1 Comparison of Convergence between Formulations

In order to compare the properties of the different formulations, and in particular their numerical stability and accuracy, we consider a manufactured solution test case with a smooth analytical solution. We define $\Omega = [0, 1]^2$ and $\Omega_i = [0.25, 0.75]^2$ with $|\Gamma_e^N| = 0$. For simplicity, let $\sigma_i = 1$, $\sigma_e = 2$, $C_m = 1$, $\Delta t \in \{1, 10^{-4}\}$ and consider the exact solution

$$u_e = \sin(\pi(x + y)),$$
$$u_i = \frac{u_e}{\sigma_i} + \cos\left(\pi(x - \tfrac{1}{4})(x - \tfrac{3}{4})\right) \cos\left(\pi(y - \tfrac{1}{4})(y - \tfrac{3}{4})\right), \qquad (5.17)$$

which corresponds to (Δt dependent) right hand sides $f = u_i - u_e - \Delta t I_m$. Note, that with (5.17) both $v \neq 0$ and $I_m \neq 0$. We discretize the domain by a uniform mesh by dividing the unit square into $n \times n$ squares and dividing each square by the (left) diagonal into isosceles triangles of size h, cf. Figure 5.1. To compare the dimensionality of the different formulations, Table 5.1 lists the dimensions of the four different finite element pairings over these meshes.

h	$|V_{e,h}|$	$|V_{i,h}|$	$|Q_h|$	$|S_h|$	$|V_h|$	$|W_h|$
4.42E-02	864	289	64	3136	2048	64
2.21E-02	3264	1089	128	12416	8192	128
1.10E-02	12672	4225	256	49408	32768	256
5.52E-03	49920	16641	512	197120	131072	512
2.76E-03	198144	66049	1024	787456	524288	1024
1.38E-03	789504	263169	2048	3147776	2097152	2048
6.93E-04	3151872	1050625	4096	12587008	8388608	4096

Table 5.1: Dimensions of the different finite element spaces for uniform refinements of the unit square. The first row corresponds to a mesh of Ω with $n = 16$, i.e. having $2 \times 16 \times 16$ cells.

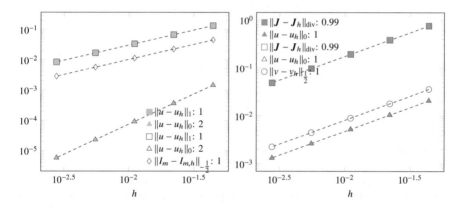

Fig. 5.2: Convergence properties of (left) primal formulations (5.6)–(5.8) and (right) mixed formulations (5.12)–(5.14). The EMI system (5.1) is solved with the exact solution given by (5.17) and $\Delta t = 1$. Filled symbols correspond to single-dimensional formulations. The number associated with each line indicates the convergence rate obtained from a least squares fit of the corresponding data.

On a series of meshes and for a set of different timesteps, we compute [2] the different approximations of all four finite element methods. We then evaluate the approximation error by evaluating the difference between (higher order interpolants of) the exact solution and the approximations in appropriate norms. Figure 5.2 shows the errors of the different formulations (with $\Delta t = 1$). In the primal formulations, u refers to the global potential, i.e. $u_i = u|_{\Omega_i}$, $u_e = u|_{\Omega_e}$. As the formulations seek the approximations $u_i \in H^1(\Omega_i)$, $u_e \in H^1(\Omega_e)$ the error is considered in the natural (broken) norm $\|u\|_1 = (\|u_i\|_1^2 + \|u_e\|_1^2)^{1/2}$. We observe that all the quantities converge linearly in their respective natural norms, as expected. In particular, the errors in the current density I_m in (5.8) and the transmembrane potential v in (5.14) are reported in the fractional norms $H^{-1/2}$ and $H^{1/2}$, respectively. The former is computed by first interpolating the error into the space of continuous piecewise cubic polynomials on Γ_h while for $v - v_h$ the P_1 element is used for error interpolation. Without this higher-order approximation of the error, i.e. if the error is computed in the same space as the discrete solution, we observe quadratic convergence.

Finally, we note that the primal formulations yield identical approximations of u cf. Figure 5.2 (left). Similarly, the mixed formulations give identical approximations of (u, J) cf. Figure 5.2 (right). Considering for comparison the error in the potential in the L^2-norm, it can be seen that the primal formulations are more accurate than the mixed formulations. The same experiments for $\Delta t = 10^{-4}$ give similar approximation results. However, it is not true that these conclusions hold in the limit of Δt approaching 0, see e.g. Chapter 6.

5.5.2 Post-Processing the Transmembrane Potential

With the exception of the multi-dimensional mixed formulation (5.14), the transmembrane potential v in the remaining EMI formulations is computed by postprocessing. In (5.6) the approximation v_h can be obtained by interpolating the difference $u_{i,h} - u_{e,h}$ onto e.g. the space of continuous piecewise linear functions over Γ_h. This procedure can, of course, be used in the other formulations as well. However (5.8) and (5.12) also offer an alternative approach. In the multi-dimensional primal formulation, the discrete membrane current density, $I_{m,h}$ is computed in the space $P_1(\Gamma_h)$ of continuous piecewise linear functions on Γ_h. In turn, v_h can be computed (in the same space) as a projection of $C_m^{-1} \Delta t I_{m,h} + f$. The same formula can be applied in the single-dimensional mixed formulation since the current density can be evaluated as $\boldsymbol{J}_h \cdot \boldsymbol{n}_i$. We recall that in (5.12) the natural space for v_h is the space of (discontinuous) piecewise constant functions on Γ_h however.

Convergence of the transmembrane potential obtained by the different formulations and post-processing strategies is shown in Figure 5.3 for the same test case as previ-

---

[2] The code used to produce results in this chapter is available at `https://github.com/MiroK/emi-book-fem` and archived at (12).

ously. We observe that the primal formulations yield quadratic convergence and that the single-dimensional primal (5.6), and multi-dimensional primal formulation (5.8) are practically identical. The discrete potentials obtained by solving the mixed formulations converge linearly with the projection method in the single-dimensional mixed formulation yielding the most accurate v_h. In particular, the approximation is better than that of the multi-dimensional mixed formulation for this test case. Computing the potential in the single-dimensional mixed formulation (5.12) by interpolating $u_{i,h} - u_{e,h}$ leads to poorer approximation than for the multi-dimensional mixed formulation. By comparing the results for two different time steps, we observe that the rates do not change considerably if Δt is modified.

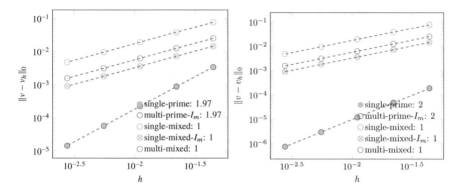

Fig. 5.3: Approximation of the transmembrane potential by the different EMI formulations. (Left) $\Delta t = 1$, (right) $\Delta t = 10^{-4}$. Postprocessing by projection (using the current density) is indicated by I_m. In multi-dimensional mixed formulation v_h is obtained by solving (5.14). Interpolation of $u_{i,h} - u_{e,h}$ is used in other formulations. The final number indicates the convergence rate.

5.6 Conclusions and Outlook

All four finite element formulations provide a converging approximation to the stationary problem (5.1). This system is a key building block in any operator splitting algorithm for the time-dependent EMI equations (1.30). The formulations provide solutions which differ by accuracy as well as computational cost, cf. Figures 5.2–5.3 and Table 5.1. The formulations also differ in robustness of their approximation properties with respect to Δt. This issue is however beyond the scope of this chapter, and the interested reader is referred to the discussion in Chapter 6. In terms of coupling with the membrane dynamics the single/multi-dimensional primal and single-dimensional mixed formulations require post-processing. However, all approaches discussed in Section 5.5.2 are easy to implement. Therefore, the choice of which formulation to employ in solving the EMI model is largely a matter of desired accuracy and available computational resources.

References

1. Babuška I (1973) The finite element method with Lagrangian multipliers. Numerische Mathematik 20(3):179–192
2. Babuška I, Gatica GN (2003) On the mixed finite element method with Lagrange multipliers. Numerical Methods for Partial Differential Equations: An International Journal 19(2):192–210
3. Boffi D, Brezzi F, Fortin M, et al. (2013) Mixed Finite Element Methods and Applications, vol 44. Springer Berlin Heidelberg, Berlin, Heidelberg
4. Chandler-Wilde SN, Hewett DP, Moiola A (2015) Interpolation of Hilbert and Sobolev spaces: quantitative estimates and counterexamples. Mathematika 61(2):414–443
5. Evans LC (2010) Partial Differential Equations, vol 19. American Mathematical Soc., Providence, Rhode Island
6. Girault V, Raviart PA (2012) Finite Element Methods for Navier-Stokes Equations: Theory and Algorithms, vol 5. Springer Berlin Heidelberg, Berlin, Heidelberg
7. Jæger KH, Tveito A (2020) Derivation of a cell-based mathematical model of excitable cells. In: Tveito A, Mardal KA, Rognes ME (eds) Modeling Excitable Tissue - The EMI Framework, Simula Springer Notes in Computing, SpringerNature
8. Jæger KH, Hustad KG, Cai X, Tveito A (2020) Operator splitting and finite difference schemes for solving the EMI model. In: Tveito A, Mardal KA, Rognes ME (eds) Modeling Excitable Tissue - The EMI Framework, Simula Springer Notes in Computing, SpringerNature
9. Könnö J, Schötzau D, Stenberg R (2011) Mixed finite element methods for problems with Robin boundary conditions. SIAM Journal on Numerical Analysis 49(1):285–308
10. Kuchta M, Mardal KA (2020) Iterative solvers for cell-based EMI models. In: Tveito A, Mardal KA, Rognes ME (eds) Modeling Excitable Tissue - The EMI Framework, Simula Springer Notes in Computing, SpringerNature
11. Kuchta M, Nordaas M, Verschaeve JCG, Mortensen M, Mardal KA (2016) Preconditioners for saddle point systems with trace constraints coupling 2D and 1D domains. SIAM Journal on Scientific Computing 38(6):B962–B987
12. Kuchta M, Mardal KA, Rognes ME (2020) Software for EMI - Solving the EMI equations using finite element methods. DOI 10.5281/zenodo.3769254, URL https://doi.org/10.5281/zenodo.3769254
13. Wohlmuth BI (2000) A mortar finite element method using dual spaces for the Lagrange multiplier. SIAM Journal on Numerical Analysis 38(3):989–1012

Chapter 6
Iterative Solvers for EMI Models

Miroslav Kuchta[1] and Kent-André Mardal[1,2]

Abstract This chapter deals with iterative solution algorithms for the four EMI formulations derived in (17, Chapter 5). Order optimal monolithic solvers robust with respect to material parameters, the number of degrees of freedom of discretization as well as the time-stepping parameter are presented and compared in terms of computational cost. Domain decomposition solver for the single-dimensional primal formulation is discussed.

6.1 Introduction

Spatial discretization of EMI models describing a few cells with a complex/realistic geometry or a large collection of cells leads to linear systems with considerable number of unknowns. In our largest simulations, we may have linear systems involving billions of unknowns at millions of time steps. It is the purpose of this chapter to discuss how such systems can be solved efficiently with available algorithms.

Let us denote the system size as N and let h be a typical grid/mesh size. The complexity of a solution algorithm can then be analyzed in terms of how the computational time grows with N. For instance, a naive Gaussian elimination would perhaps scale as $O(N^3)$ and for linear system involving billions of unknowns such an approach is therefore not feasible during a life-time. Here, we shall aim for algorithms that are *order optimal*, i.e. their solution time scales linearly, $O(N)$, with respect to the number of unknowns. Clearly, $O(N)$ is order optimal in the sense that this is also the complexity of writing the results to file. In general, direct solvers, like for instance LU, do not scale linearly in N. Here, we will use Krylov solvers like the Conjugate

---

[1] Simula Research Laboratory, Norway
[2] Department of Mathematics, University of Oslo, Oslo, Norway

The Author(s) 2021
A. Tveito et al. (eds.), *Modeling Excitable Tissue*, Simula SpringerBriefs
on Computing 7, https://doi.org/10.1007/978-3-030-61157-6_6

Gradient (CG) or the Minimal Residual (MinRes) methods. Both methods are known to provide efficient computations when combined with proper preconditioning techniques. Preconditioners based on multigrid and/or domain decomposition methods have been shown to be order-optimal for a variety of problems, but the theory for the EMI problem is currently limited. Efficient solvers for the EMI-problem are discussed in the following.

We say that a solver for the *transient* EMI model (1.30) is order optimal if the solution time grows linearly in Δt^{-1}, where Δt is the time step. Considering the *stationary* problem (5.1), which is to be solved at every step of the temporal loop, it is clear that order optimality of the transient solver requires that the solver of the linear system does not degenerate for small (or large) Δt, i.e. that it is robust with respect to the time stepping.

In this chapter we discuss solution algorithms for the linear systems due to the finite element discretization of (5.1) which are robust in h and well as Δt. Two types of approaches will be considered. Monolithic approaches, where all the unknowns are solved for at once, are the subject of Section 6.2. Section 6.3 then concerns a domain decomposition approach where one iterates between the intra/extra-cellular subproblems. The solvers will be compared in terms of robustness and cost, however, only serial performance will be addressed. We remark that parallel scalable solvers suitable for the mixed formulations of the EMI models (5.12) (5.14) are the balancing domain decomposition methods, see e.g. (24; 25). For the elliptic single-dimensional primal formulation (5.6), the FETI domain decomposition methods, e.g. (13; 18) could be used. Moreover, to simplify the exposition the focus shall be on a single cell model. We remark that all the algorithms presented further can be generalized in a rather straightforward manner to models with multiple *disconnected* cells. However, construction of robust monolithic solvers for *multi-dimensional* formulations for collections of *connected* cells is out of the scope of this manuscript.

6.2 Monolithic Solvers

Let $\mathcal{A}_h x_h = L_h$ be a linear system due to discretization of the continuous problem $\mathcal{A}x = L$ in W', where W' is a dual space to some Hilbert space W and $\mathcal{A} : W \to W'$. Note that in case of (5.1) the continuous operator $\mathcal{A}$ depends on material parameters σ_i, σ_e, C_m as well as the time step size Δt. Here the focus is placed on the latter dependence.

The monolithic solvers considered here are preconditioned Krylov methods where the preconditioner $\mathcal{B} : W' \to W$ shall be constructed such that the number of iterations required for solving problems $\mathcal{B}_h \mathcal{A}_h x_h = \mathcal{B}_h L_h$ is bounded in h and Δt. A constructive framework for establishing $\mathcal{B}$ is operator preconditioning (19), where the structure of the preconditioner reflects mapping properties of $\mathcal{A}$ as an isomorphism between suitably chosen Hilbert space and its dual space. In particular, if $\mathcal{A}: W \rightarrow W'$ is an isomorphism then a Riesz map with respect to the inner product of W is a suitable preconditioner. As explained in (19), multilevel or domain decomposition algorithms are equivalent to Riesz maps for standard elliptic problems. Furthermore, these schemes also work in a fractional setting (1); a property which will be exploited in this chapter. We shall illustrate the framework briefly below using the single-dimensional formulations (5.6), (5.12) as an example.

In the following we follow the notation of Chapter 5. In particular, $\|\cdot\|_{0,\Omega}$ denotes the L^2 norm of a scalar or vector field over Ω, while $\|\cdot\|_{1,\Omega}$, $\|\cdot\|_{\text{div},\Omega}$ are the $H^1(\Omega)$ and $\boldsymbol{H}(\text{div},\Omega)$ norms respectively. For a Hilbert space W other than L^2, H^1, $\boldsymbol{H}(\text{div})$ the norm is denoted as $\|\cdot\|_W$ while $(\cdot,\cdot)_W$ denotes the inner product on W. Moreover, we let $(\cdot,\cdot)_\Omega$ be the L^2 inner product. If the domain is clear the subscript will be omitted. By $(\cdot,\cdot)$ we shall also denote a duality pairing between a Hilbert space and its dual. The dual of an operator A with respect to the L^2 inner product is then denoted as A'. Finally, let us introduce scaled, sum and intersection spaces, which will be required for well-posedness of some of the formulations. If W, Q are Hilbert spaces and a an arbitrary positive real number, the scaled space aW, the sum space $W + Q$ and the intersection space $W \cap Q$ are Hilbert spaces with norms $\|x\|_{aW} = a\|\cdot\|_W$,

$$\|x\|_{W+Q} = \inf_{\substack{w+q=x \\ w\in W, q\in Q}} \sqrt{\|w\|_W^2 + \|q\|_Q^2} \quad \text{and} \quad \|x\|_{W\cap Q} = \sqrt{\|x\|_W^2 + \|x\|_Q^2}.$$

6.2.1 Single-Dimensional Primal Solvers

In order to simplify the analysis let σ_i, σ_e and C_m be positive constants and let us consider a slightly modified (cf. the underlined term) single-dimensional primal formulation (5.6): Find $u_i \in V_i = H^1(\Omega_i)$, $u_e \in V_e = H^1_{\Gamma^D_e}(\Omega_e)$ such that for all $v_e \in V_e$, $v_i \in V_i$

$$(\sigma_e \nabla u_e, \nabla v_e) + \frac{C_m}{\Delta t}(T_e u_e, T_e v_e) - \frac{C_m}{\Delta t}(T_i u_i, T_e v_e) = \frac{C_m}{\Delta t}(f, v_e),$$
$$-\frac{C_m}{\Delta t}(T_e u_e, T_i v_i) + (\sigma_i \nabla u_i, \nabla v_i) + \underline{(u_i, v_i)} + \frac{C_m}{\Delta t}(T_i u_i, T_i v_i) = -\frac{C_m}{\Delta t}(f, v_i). \tag{6.1}$$

Here, T_e, T_i are the trace operators $T_i u_i = u_i|_\Gamma$ and $T_e u_e = u_e|_\Gamma$. Letting $W = V_e \times V_i$, $u = (u_e, u_i)$ the problem (6.1) can be stated as $\mathcal{A}u = L$ in W' with the operator $\mathcal{A}$ and functional L defined as

$$\mathcal{A} = \begin{pmatrix} -\nabla\cdot(\sigma_e\nabla) + \frac{C_m}{\Delta t}T'_e T_e & -\frac{C_m}{\Delta t}T'_e T_i \\ -\frac{C_m}{\Delta t}T'_i T_e & -\nabla\cdot(\sigma_i\nabla) + I + \frac{C_m}{\Delta t}T'_i T_i \end{pmatrix} \quad \text{and} \quad L(v) = \frac{C_m}{\Delta t}\begin{pmatrix} (f, v_e) \\ -(f, v_i) \end{pmatrix}. \tag{6.2}$$

Note that the underlined lower order term in (6.1) corresponds to I in $\mathcal{A}$.

To precondition (6.1) using the operator preconditioning framework we proceed to show that $\mathcal{A} : W \to W'$ is an isomorphism. This statement follows from the Lax-Milgram theorem, see e.g. (6, Ch 2.7). Let $\|\cdot\|_W$ be *some* norm of the space W (we shall see shortly that different norms can be considered leading to different preconditioners). Hence coercivity of $\mathcal{A}$, i.e.

$$\text{There exists } \alpha_* > 0 \text{ such that } \alpha_*\|u\|_W^2 \le (\mathcal{A}u, u), \quad \forall u \in W \qquad (6.3)$$

and boundedness of $\mathcal{A}$, i.e.

$$\text{There exists } \alpha^* > 0 \text{ such that } (\mathcal{A}u, v) \le \alpha^*\|u\|_W\|v\|_W, \quad \forall u, v \in W \qquad (6.4)$$

need to be shown. While the details of the proof are beyond the scope of the current text, we remark that using the space $W = H^1_{\Gamma_e^D}(\Omega_e) \times H^1(\Omega_i)$, we obtained the bound

$$(\mathcal{A}u, u) \le \max\left(\sigma_e + 2C_e\frac{C_m}{\Delta t}, \sigma_i, 1, 2C_i\frac{C_m}{\Delta t}\right)\|u\|_W^2 \qquad (6.5)$$

by trace, Cauchy-Schwarz, and Young inequalities. Hence, the operator $\mathcal{A}$ is bounded. Note, however, that the constant α^* depends on Δt and, in particular, it blows up for small time steps. Then, the lower bound is

$$\begin{aligned}(\mathcal{A}u, u) &= \sigma_e\|\nabla u_e\|_0^2 + \sigma_i\|\nabla u_i\|_0^2 + \|u_i\|_0^2 + \frac{C_m}{\Delta t}\|T_e u_e - T_i u_i\|_0^2 \\ &= \min(1, \sigma_i, \sigma_e)\|u\|_W^2\end{aligned} \qquad (6.6)$$

and the coercivity thus holds with constant $\alpha_* = \min(1, \sigma_i, \sigma_e)$ which is independent of the time step. The condition number of the preconditioned system, using a preconditioner deduced from W, is then $\frac{\alpha^*}{\alpha_*}$ and from this theoretical consideration we may then expect the number of iterations to grow linearly as Δt decreases.

The exact preconditioner based on W is block diagonal opearator

$$\mathcal{B} = \begin{pmatrix} -\nabla\cdot(\sigma_i\nabla) & 0 \\ 0 & -\nabla\cdot(\sigma_e\nabla) + I \end{pmatrix}^{-1}. \qquad (6.7)$$

Observe that the preconditioner is efficient in the sense that its evaluation mounts to solving two smaller subproblems posed on Ω_i and Ω_e respectively. However, from our analysis we expect the performance of $\mathcal{B}$ to deteriorate for small time steps. Indeed, this behavior is confirmed by results[1] in Table 6.1. Therein it can be seen

---

[1] All the experiments in this chapter are conducted using the manufactured problem (5.17) considered in the convergence study in Chapter 5. In particular, we use uniform meshes obtained by dividing the unit square ($\Omega_e \cup \Omega_i$) first into n_h^2 squares each of which is then subdivided to 2 isosceles triangles with diameter h. The finite element discretization is as described in Chapter 5. We set $\sigma_i = 1$, $\sigma_e = 2.2$ and $C_m = 1$.

The discrete linear systems $\mathcal{A}_h x_h = L_h$ are solved iteratively using the CG or MinRes methods. Implementation of the methods is provided by (2). The solvers are always started from a random initial vector. As a convergence criterion the relative preconditioned residual norm is required to

that the iterations stabilize with h for $\Delta t > 10^{-4}$ while for smaller time steps finer meshes are needed to attain the bounds.

In an attempt to find a more robust preconditioner let us observe that estimates (6.5), (6.6) show that $u \mapsto (\mathcal{A}u, u)^{1/2}$ defines a different norm on W. Let now $\|u\|_W = (\mathcal{A}u, u)^{1/2}$. Then $(\mathcal{A}u, u) = \|u\|_W^2$ and $(\mathcal{A}u, v) \leq (\mathcal{A}u, u)^{1/2}(\mathcal{A}v, v)^{1/2}$ so that the conditions (6.3) and (6.4) hold with $\alpha^* = 1$, $\alpha_* = 1$. In another words, a Riesz map preconditioner with respect to the inner product $(u, v)_W = (\mathcal{A}u, v)$ is independent of Δt. As the Riesz map in this case is in fact $\mathcal{A}^{-1}$ its exact evaluation (by LU) is not feasible. However, for the purpose of preconditioning, it suffices to replace the mapping by a spectrally equivalent operator. Then, provided that the equivalence is robust in Δt the approximate operator will lead to iterations bounded in time step and the discretization parameter. We remark that since the preconditioner takes the entire $\mathcal{A}$ into account, including the off-diagonal terms cf. block diagonal operator (6.7), we shall refer to it as monolithic.

n_h Δt	Diagonal (6.7)						Monolithic BoomerAMG($\mathcal{A}$)					
	2^3	2^4	2^5	2^6	2^7	2^8	2^3	2^4	2^5	2^6	2^7	2^8
10^8	8	5	5	5	5	5	8	8	8	8	8	9
10^6	7	7	7	7	7	7	8	8	8	8	8	9
10^4	7	7	7	7	7	7	8	8	8	8	8	9
10^2	6	6	6	6	6	6	7	8	8	8	8	9
1	11	11	11	11	10	10	8	8	8	9	9	9
10^{-2}	26	30	33	32	33	32	8	8	9	9	10	10
10^{-4}	39	70	90	114	139	155	28	26	19	14	11	10
10^{-6}	42	79	134	181	234	300	56	81	101	104	87	62
10^{-8}	42	78	130	179	244	327	81	140	226	320	359	378

Table 6.1: Primal single-dimensional formulation. Number of CG iterations using (left) the diagonal operator (6.7) and (right) the Riesz map with respect to the $\mathcal{A}$ induced norm. Neither preconditioner is robust for $\Delta t \ll 1$.

Since (5.6) leads to symmetric positive-definite matrices, cf. the coercivity condition (6.3), we shall here use algebraic multigrid to construct the approximation of $\mathcal{A}^{-1}$. More precisely, the action of the operator is realized by single V-cycle of BoomerAMG (23). In Table 6.1 it can be seen that this choice leads to h bounded CG iterations for $\Delta t > 10^{-8}$. However, there is a clear sensitivity of the bound to Δt. Thus BoomerAMG approximations of $\mathcal{A}^{-1}$ are not Δt- robust. In fact, estimates of the condition number for the preconditioned problems with $n_h = 2^8$ are 1.2, 1.5, 2.0, 9.2 and 53 for $\Delta t = 1, 10^{-2}, 10^{-4}, 10^{-6}$ and 10^{-8} respectively.

---

be less than 10^{-12} in magnitude. Unless specified otherwise the preconditioners $\mathcal{B}_h$ are evaluated exactly by LU. Finally, condition number estimates of the preconditioned linear systems are obtained by using iterative Krylov-Schur solver from (10) applied to the generalized eigenvalue problem $\mathcal{A}_h x_h = \lambda_h \mathcal{B}_h^{-1} x_h$. The reported condition number is then $\max|\lambda_h|/\min|\lambda_h|$.

The source code used for the experiments can be found on https://github.com/MiroK/emi-book-solvers and is arxived at (15).

6.2.2 Single-Dimensional Mixed Solvers

We next consider preconditioners for the single-dimensional mixed formulation (5.12): Find $E \in S$, $u \in Q$ such that

$$(\sigma^{-1}J, \tau) + \frac{\Delta t}{C_m}(T_n J, T_n \tau) - (u, \nabla \cdot \tau) = -(f, T_n \tau), \quad \forall \tau \in S,$$
$$(q, \nabla \cdot J) = 0, \quad \forall q \in Q,$$

where $S = \boldsymbol{H}_{\Gamma_e^N}(\text{div}, \Omega)$, $Q = L^2(\Omega)$ and T_n is the normal trace operator $T_n \tau = \tau \cdot n_i$. Letting $W = S \times Q$, $x = (J, u)$ the single-dimensional mixed formulation is equivalent to the problem $\mathcal{A}x = L$ in W' with

$$\mathcal{A} = \begin{pmatrix} \sigma^{-1}I + \frac{\Delta t}{C_m}T_n'T_n & -\nabla \\ \nabla \cdot & 0 \end{pmatrix} \quad \text{and} \quad L(x) = -(f, T_n J). \tag{6.8}$$

Note that in (6.8) the membrane term $(T_n J, T_n \tau)$ is weighted by $\Delta t / C_m$, cf. (6.2). Thus, unlike in the single-dimensional primal formulation, the term does not dominate for small time steps.

In order to apply the operator preconditioning framework to (6.8) the operator $\mathcal{A}$ needs to be shown to be an isomorphism. To this end we consider $\mathcal{A}$ as an abstract operator over $W = S \times Q$ with the form

$$\mathcal{A} = \begin{pmatrix} A & B' \\ B & 0 \end{pmatrix} \quad \text{where} \quad \begin{array}{l} A : S \to V', \\ B : S \to Q', \end{array} \tag{6.9}$$

and apply the Brezzi theory (7). This leads us to the potential preconditioners for the multi-dimensional mixed problem

$$\mathcal{B}_1 = \begin{pmatrix} \sigma^{-1}I - \nabla \nabla \cdot & 0 \\ 0 & I \end{pmatrix}^{-1} \quad \text{and} \quad \mathcal{B}_2 = \begin{pmatrix} \sigma^{-1}I - \nabla \nabla \cdot + \frac{\Delta t}{C_m}T_n'T_n & 0 \\ 0 & I \end{pmatrix}^{-1}, \tag{6.10}$$

which are the Riesz mappings with respect to the inner products of the spaces

$$W_1 = S \times Q \quad \text{and} \quad W_2 = S \cap \sqrt{\frac{\Delta t}{C_m}}N \times Q, \tag{6.11}$$

where $N = \{v \in S; \|v \cdot n\|_{0,\Gamma} < \infty\}$. Note that in W_2 we enforce additional regularity on the normal trace since T_n maps S to $H^{-1/2}(\Gamma)$, e.g. (9, Ch. 2). Moreover the traces are considered in a weighted space.

A consequence of the mapping properties of the normal trace operator is the fact that the term $(T_n J, T_n \tau)$ cannot be bounded using the $\boldsymbol{H}(\text{div})$ norm and in turn the Brezzi conditions are not satisfied on W_1. We remark that if such a bound were possible the boundedness constant would depend on Δt, cf. (6.5). As the Brezzi conditions do not hold on W_1 we expect preconditioner $\mathcal{B}_1$ to perform poorly. Indeed, Table 6.2 shows that the condition numbers of $\mathcal{B}_{1,h}\mathcal{A}_h$ are not bounded in h. The condition numbers for fixed h can be seen to grow with the time step. In Table 6.3 the ill-posedness of the problem in W_1 manifests as unstable iterations or failure of MinRes to converge. We note that for small Δt the iterations appear bounded in h.

n_h / Δt	$\mathcal{B}_1$						$\mathcal{B}_2$					
	2^2	2^3	2^4	2^5	2^6	2^7	2^2	2^3	2^4	2^5	2^6	2^7
10^2	4223	7540	14766	29437	4785	9569	13.52	13.52	13.52	13.52	13.52	13.52
10^1	75.0	132	259	515	1028	2057	2.28	2.28	2.28	2.28	2.28	2.28
1	8.86	15	27	52	104	207	2.20	2.20	2.20	2.20	2.20	2.20
10^{-1}	2.27	2.85	4.11	6.66	12	22	2.20	2.20	2.20	2.20	2.20	2.20
10^{-2}	2.20	2.20	2.20	2.20	2.59	3.60	2.20	2.20	2.20	2.20	2.20	2.20

Table 6.2: Conditioning of single-dimensional mixed formulation. (Left) Posing the problem in $H(\mathrm{div}) \times L^2$ violates Brezzi conditions. (Right) Preconditioner based on W_2 in (6.11) is inf-sup stable with the inf-sup constant depending on Δt for $\Delta t / C_m > 1$.

Posing (6.8) in W_2 it can be shown that the Brezzi conditions hold with the constants independent of the time step.

We shall not prove the validity of the Brezzi theory here. Instead, Table 6.2 offers numerical evidence that the discrete condition is satisfied. Therein condition numbers of the $\mathcal{B}_2$ preconditioned problems are reported and boundedness in h can be observed. Moreover, it can be seen that the inf-sup constant depends on Δt, in particular, the bound goes to 0 as Δt grows. However, on the subspace $\{(J, u) \in W_2; (u, 1)_{0,\Omega_i} = 0\}$ the inf-sup condition holds independent of Δt. The single run-away mode appears to have no effect on the MinRes iterations, cf. Table 6.3, see also (20).

n_h / Δt	$\mathcal{B}_1$						$\mathcal{B}_2$					
	2^3	2^4	2^5	2^6	2^7	2^8	2^3	2^4	2^5	2^6	2^7	2^8
10^8	–	–	–	–	–	–	10	10	10	10	10	10
10^6	–	–	–	–	–	–	15	12	12	12	12	12
10^4	–	–	–	–	–	–	15	15	15	16	16	16
10^2	156	–	–	–	–	–	15	15	15	15	16	16
10^1	26	36	57	86	129	209	14	14	14	14	14	14
1	35	50	72	107	155	234	16	16	16	16	16	16
10^{-1}	16	19	23	32	44	60	16	17	17	18	18	18
10^{-2}	18	19	21	27	34	45	19	19	19	19	20	20
10^{-4}	19	19	20	20	20	20	19	19	20	20	20	21
10^{-6}	19	19	20	20	21	21	19	19	20	20	21	21
10^{-8}	19	19	20	20	21	21	19	19	20	20	21	21

Table 6.3: MinRes iterations with preconditioned single-dimensional mixed formulation using preconditioners (6.10). Failure to converge within 500 iterations is denoted by –. Convergence of $\mathcal{B}_2$ seems unaffected by the single Δt unbounded inf-sup mode.

Considering order optimality of the $\mathcal{B}_2$ based solver we recall that the iterations in Table 6.3 were run with the exact preconditioner. In particular, the $S \cap N$ Riesz map was computed by LU. In practical computations such realization is not feasible and scalable approximation is needed. For standard $\boldsymbol{H}(\text{div})$ inner product, i.e. the one used in $\mathcal{B}_1$, the Riesz map can be efficiently realized by multigrid algorithms (14). However, in the authors' experience this approach does not work equally well with the $S \cap N$ inner product.

We have seen in Sections 6.2.1 and 6.2.2 that construction of order optimal monolithic solvers for the single-dimensional EMI formulations presents a challenge. In primal formulation Δt robustness of the monolithic approach was problematic. For the mixed formulation, order optimality required a specialized solver for the Riesz mapping over the subspace of $\boldsymbol{H}(\text{div})$. These two problems are addressed by solvers for multi-dimensional formulations.

6.2.3 Multi-Dimensional Solvers

In this section we consider the construction of preconditioners for the operators

$$\mathcal{A}_p = \begin{pmatrix} -\nabla\cdot(\sigma_e\nabla) & 0 & -T_e' \\ 0 & -\nabla\cdot(\sigma_i\nabla) & T_i' \\ -T_e & T_i & -\frac{\Delta t}{C_m}I \end{pmatrix} \quad \text{and} \quad \mathcal{A}_m = \begin{pmatrix} \sigma^{-1}I & -\nabla & T_n' \\ \nabla\cdot & 0 & 0 \\ T_n & 0 & -\frac{C_m}{\Delta t}I \end{pmatrix}, \quad (6.12)$$

which induce respectively the multi-dimensional primal weak formulation (5.8) and the multi-dimensional mixed weak formulation (5.14). In order to discuss well-posedness let

$$W_p = H^1_{\Gamma_e^D}(\Omega_e) \times H^1(\Omega_i) \times H^{-1/2}(\Gamma) \cap \sqrt{\tfrac{\Delta t}{C_m}}L^2(\Gamma) \quad (6.13)$$

and

$$W_m = \boldsymbol{H}_{\Gamma_e^N}(\text{div},\Omega) \times L^2(\Omega) \times H^{1/2}(\Gamma) \cap \sqrt{\tfrac{C_m}{\Delta t}}L^2(\Gamma). \quad (6.14)$$

We remark that the fractional space $H^{-1/2}$, in which the membrane current density I_m in (5.6) is sought, and $H^{1/2}$, which is the space of the transmembrane potential v in (5.14), are dictated by the mapping properties of the trace operators T_e, T_i and T_n. For example, as $T_i : H^1(\Omega_i) \to H^{1/2}(\Gamma)$, $I_m \in (H^{1/2})'$ so that the term $(T_i u_i, I_m)$ is bounded. Note also that only the spaces involving Γ are now Δt-dependent, cf. e.g. W_2 in (6.11).

Due to the Δt-weighted (penalty) terms in $\mathcal{A}_p$ and $\mathcal{A}_m$ the operators cannot be established as isomorphisms on W_p and W_m by straightforward application of the Brezzi theory. Instead, $\mathcal{A}_p$ for $\Delta t/C_m \leq 1$ and $\mathcal{A}_m$ for $\Delta t/C_m \geq 1$ can be analyzed using the framework for saddle point systems with penalty, see (4, Ch. 3.4). A crucial assumption then is that the system without the penalty term satisfies the Brezzi conditions. While here the well-posedness shall be demonstrated only by numerical experiments, let us point out an important difference between the operators. To this end let $c > 0$ be a constant and consider a piecewise constant potential u such that $u_i = c, u_e = 0$ and set $\boldsymbol{J} = \boldsymbol{0}$. Then for any $\tau \in \boldsymbol{H}(\text{div})$ by divergence theorem

$$(\nabla \cdot \tau, u)_\Omega + (T_n \tau, v) = (\nabla \cdot \tau, u)_{\Omega_i} + (T_n \tau, v) = c(T_n \tau, 1) + (T_n \tau, v)$$

and in turn, if $v = -c$ we have that

$$\begin{pmatrix} \sigma^{-1} I - \nabla T'_n \\ \nabla \cdot \quad 0 \quad 0 \\ T_n \quad 0 \quad 0 \end{pmatrix} \begin{pmatrix} \boldsymbol{J} \\ u, \\ v \end{pmatrix} = \begin{pmatrix} c(T_n \tau, 1) + (T_n \tau, v) \\ 0 \\ 0 \end{pmatrix} = \begin{pmatrix} \boldsymbol{0} \\ 0 \\ 0 \end{pmatrix}.$$

Thus the operator $\mathcal{A}_m$ without the penalty term is singular with a one dimensional kernel spanned by vector $(\boldsymbol{0}, u, -c)$ and the Brezzi conditions do not hold. We shall see implications of this property on the convergence of the iterative solvers shortly. We remark that $\mathcal{A}_p$ without the penalty term is non-singular.

Assuming that $\mathcal{A}_p : W_p \to W'_p$ is an isomorphism we consider as a preconditioner for the single-dimensional primal formulation the operator

$$\mathcal{B}_p = \begin{pmatrix} -\nabla \cdot (\sigma_e \nabla) & 0 & 0 \\ 0 & -\nabla \cdot (\sigma_i \nabla) + I & 0 \\ 0 & 0 & (-\Delta + I)^{-1/2} + \frac{\Delta t}{C_m} I \end{pmatrix}^{-1}. \quad (6.15)$$

Observe that the preconditioner consists of (approximate) solvers for three decoupled subproblems posed on Ω_e, Ω_i and Γ. Moreover, the problems on extra and intracellular domains are standard elliptic operators for which efficient (black-box) multigrid techniques exist, e.g. (23). The problem on the interface is then less standard as it concerns a fractional Helmholtz operator. However, efficient multilevel solvers have been established e.g. in (5; 1). We remark that if the discrete spaces for I_m (or v) contain only a few thousands of degrees of freedom an eigenvalue realization of the fractional preconditioner is feasible cf. (16). As the interfacial spaces here are small, cf. Table (5.1), we use further the spectral approach.

n_h / Δt	2^2	2^3	2^4	2^5	2^6	2^7
10^2	2752	2752	2752	2752	2752	2752
10^1	29.61	29.62	29.62	29.62	29.62	29.62
1	6.86	6.88	6.89	6.89	6.89	6.89
10^{-1}	8.84	8.88	8.89	8.89	8.89	8.89
10^{-2}	8.89	8.93	8.94	8.94	8.94	8.94

n_h / Δt	2^4	2^5	2^6	2^7	2^8	2^9
10^8	10	10	10	10	10	10
10^6	10	10	10	10	10	10
10^4	15	15	15	15	15	15
10^2	18	18	18	17	17	17
1	27	27	27	26	26	26
10^{-2}	43	47	46	44	44	43
10^{-4}	49	60	62	60	57	56
10^{-6}	51	60	62	61	59	59
10^{-8}	49	60	62	61	59	59

Table 6.4: Fractional preconditioner (6.15) for multi-dimensional primal formulation. (Left) Condition numbers are bounded in h. Growth in Δt for $\Delta t > 1$ is caused by a single mode. (Right) MinRes iterations counts are bounded in h and time step.

Using (6.15) Table 6.4 reports the condition numbers of the preconditioned problems $\mathcal{B}_{p,h}\mathcal{A}_{p,h}$. It can be seen that the results are bounded in h, thus providing a numerical evidence for $\mathcal{A}_p$ being an isomorphism on W_p in (6.13). Moreover, it can be seen that the conditioning of the problem deteriorates with Δt large. However, in this case it is only a single mode $u_i = 0$, $u_e = \text{const}$, $I_m = 0$ which causes the blowup. Considering the MinRes iteration counts, the presence of this mode seems to have almost no impact on boundedness in h or Δt.

We finally come back to the multi-dimensional mixed formulation. Based on the space W_m in (6.14) let a preconditioner for the multi-dimensional mixed formulation be

$$
\mathcal{B}_m = \begin{pmatrix} \sigma^{-1}I - \nabla \nabla \cdot & 0 & 0 \\ 0 & I & 0 \\ 0 & 0 & (-\Delta + I)^{1/2} + \frac{C_m}{\Delta t} I \end{pmatrix}^{-1}. \tag{6.16}
$$

We remark that the preconditioner uses a standard $\boldsymbol{H}(\mathrm{div})$ inner product, cf. $\mathcal{B}_2$ in (6.10), which can be efficiently realized by multigrid methods, e.g. (14). Note also that the fractionality of the Laplacian is $1/2$, cf. $-1/2$ of the multi-dimensional primal preconditioner (6.15). In addition to the previously mentioned multilevel methods, problems $(-\Delta u + u)^s x = b$ for $0 < s < 1$ can be efficiently solved by a number of approaches, see (3) and references therein.

Using (6.16) Table 6.5 reports the condition numbers of the preconditioned multi-dimensional mixed formulation. The conditioning can be seen to be stable in h, while in agreement with the limit singularity property, there is a growth with Δt. Given that only a single mode is responsible for the lack of Δt-boundedness and recalling results of Table 6.4 or 6.3 we might expect that also here the MinRes solver will not be affected. However, Table 6.6 shows that this is not the case. In fact, as Δt grows the iterations become unstable in h. Figure 6.1 then shows a typical convergence behavior of the solver. We see that the relative preconditioned residual norm is quickly reduced to about 10^{-9} in approximately 30 iterations (regardless of the mesh resolution). Afterwards the convergence stalls. We conclude that for robust preconditioning the nullspace of $\mathcal{A}_m$ must be addressed.

Δt \ n_h	$\mathcal{B}_m\mathcal{A}_m$						$\mathcal{B}_m^0\mathcal{A}_m^0$					
	2^2	2^3	2^4	2^5	2^6	2^7	2^2	2^3	2^4	2^5	2^6	2^7
10^2	17566	17627	17733	17810	17857	17882	4.40	4.42	4.51	4.58	4.63	4.67
10^1	177	178	179	180	180	180	4.39	4.41	4.50	4.57	4.62	4.66
1	3.31	3.73	3.96	4.09	4.18	4.27	3.31	3.73	3.96	4.09	4.18	4.27
10^{-1}	1.08	1.16	1.30	1.56	1.96	2.48	2.61	2.61	2.61	2.61	2.64	3.14
10^{-2}	1.03	1.03	1.03	1.03	1.03	1.03	2.62	2.62	2.62	2.62	2.62	2.62

Table 6.5: Conditioning of the multi-dimensional mixed formulation. (Left) Problem considered on W_m in (6.14) yields operator $\mathcal{A}_m$ with a singular limit as $\Delta t \to \infty$. (Right) Formulation including an additional constraint on the transmembrane potential $\int_\Gamma v \, dS = 0$ yields h and Δt robust condition numbers.

Motivated by the observation that $\mathcal{A}_m$ becomes singular in the limit $\Delta t = \infty$ with a mode $\boldsymbol{J} = \boldsymbol{0}$, $u_e = 0$, $u_i = c$, $v = -c$ in the kernel, let $W_e^0 = W_e \times \mathbb{R}$ and let us define a new solution operator for the multi-dimensional mixed formulation $\mathcal{A}_m^0 : W_e^0 \to (W_e^0)'$ and its preconditioner $\mathcal{B}_m^0 : (W_e^0)' \to W_e^0$ as

$$
\mathcal{A}_m^0 = \begin{pmatrix} \sigma^{-1}I - \nabla & T_n' & 0 \\ \nabla \cdot & 0 & 0 & 0 \\ T_n & 0 & -\frac{C_m}{\Delta t}I & I \\ 0 & 0 & I & 0 \end{pmatrix}, \mathcal{B}_m^0 = \begin{pmatrix} \sigma^{-1}I - \nabla \, \nabla \cdot & 0 & 0 & 0 \\ 0 & I & 0 & 0 \\ 0 & 0 & (-\Delta + I)^{1/2} + \frac{C_m}{\Delta t}I & 0 \\ 0 & 0 & 0 & \mu I \end{pmatrix}^{-1},
$$
(6.17)

where $\mu = \min(1, C_m/\Delta t)$. Note that $\mathcal{A}_m^0$ includes an additional unknown, a single scalar, which enforces $(1, v)_\Gamma = 0$. The new constraint thus eliminates constant transmembrane potential and in turn the new operator is non-singular[2]. Considering results reported in Tables 6.5, 6.6 it can be seen that the new preconditioned formulation leads to condition numbers and iteration counts, which are bounded in both the mesh size and the time step. Note that the extra unknown has only a small impact on the number of iterations compared to the $\mathcal{B}_m$ preconditioner.

Δt \ n_h	$\mathcal{B}_m \mathcal{A}_m$						$\mathcal{B}_m^0 \mathcal{A}_m^0$					
	2^3	2^4	2^5	2^6	2^7	2^8	2^3	2^4	2^5	2^6	2^7	2^8
10^8	457	–	241	–	–	–	28	32	34	36	37	37
10^6	54	174	82	301	258	227	28	32	34	36	37	37
10^4	37	41	45	45	47	47	28	32	34	36	37	37
10^2	32	37	39	39	41	41	28	32	34	36	37	37
1	28	30	32	34	34	35	30	33	35	35	37	38
10^{-2}	19	22	25	29	33	35	20	23	29	33	35	39
10^{-4}	10	10	10	11	11	12	12	12	12	13	13	14
10^{-6}	7	7	9	9	9	9	9	9	10	10	10	10
10^{-8}	7	7	7	7	7	7	9	9	9	9	9	9

Table 6.6: MinRes iterations for the multi-dimensional mixed formulation. No convergence in 500 iterations is indicated by –. (Left) For large Δt the iterations are unstable since operator $\mathcal{A}_m$ becomes singular in the limit as $\Delta t \to \infty$. (Right) Constraining transmembrane potential the operator $\mathcal{A}_m^0$ does not have the limit singular property.

---

[2] A physical motivation for the constraint can be found in considering the EMI interface equation $C_m \frac{\partial v}{\partial t} = I_m + I_{ion}$ on Γ. By integrating the left-hand side we obtain $C_m \frac{d}{dt}(1, v)_\Gamma$. Recall that $I_m = \sigma_i \nabla u_i \cdot \boldsymbol{n}_i$ on Γ and $-\nabla \cdot (\sigma_i \nabla u_i) = 0$ in Ω_i. Then, setting $I_{ion} = 0$ and integrating the right-hand side we have

$$(1, I_m + I_{ion})_\Gamma = (1, I_m)_\Gamma = (1, -\nabla \cdot (\sigma_i \nabla u_i))_{\Omega_i} = 0.$$

Thus we obtain a conservation relation $C_m \frac{d}{dt}(1, v)_\Gamma = 0$.

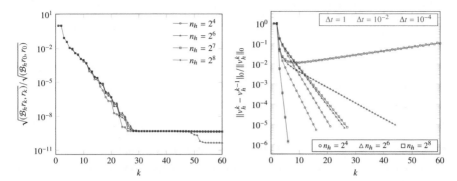

Fig. 6.1: (Left) Convergence of the MinRes method for (6.12)-based multi-dimensional mixed formulation with $\Delta t = 10^{-8}$ and problem from Table 6.6. For all mesh resolutions the norm is reduced in cca. 30 iterations. Afterwards the reduction stalls. (Right) Convergence of the Robin/Neumann domain decomposition algorithm. For small Δt the algorithm becomes sensitive to mesh size and can even diverge.

6.3 Domain Decomposition Solvers

Having seen that monolithic solvers for the EMI equations can be sensitive to spatial and temporal resolution we next briefly discuss robustness of the non-overlapping domain decomposition (DD) approach. The focus here shall only be on the single-dimensional primal formulation and the Robin/Neumann DD algorithm in the form presented in (12, Chapter 4). In particular, our implementation shall not include any coarse space, cf. (21), or a preconditioner, see e.g. (8) for interpretation of DD as Steklov-Poincaré operators in fractional Sobolev spaces.

For the sake of self-containedness we review the variational formulation of the Robin/Neumann algorithm. Let $V_e = H^1_{\Gamma_e^D}(\Omega_e)$, $V_i = H^1(\Omega_i)$ and v^0, u_e^0 be the given initial transmembrane and extracellular potentials. A single iteration of the DD algorithm then produces a new approximation v^1 in three steps. (i) We find $u_i \in V_i$ such that

$$(\sigma_i \nabla u_i, \nabla v_i) + \frac{C_m}{\Delta t}(T_i u_i, T_i v_i) = -\frac{C_m}{\Delta t}(f, v_i) + \frac{C_m}{\Delta t}(T_e u_e^0, T_i v_i), \quad \forall v_i \in V_i.$$

(ii) Using the computed intracellular potential the new extracellular potential $u_e \in V_e$ is found such that

$$(\sigma_e \nabla u_e, \nabla v_e) = (-\sigma_i \nabla u_i \cdot n_i, T_e v_e), \quad \forall v_e \in V_e.$$

(iii) Finally $v^1 = T_i u_i - T_e u_e$. The iterations continue by assigning $v^0 = v^1$ and $u_e^0 = u_e$ until convergence. which in the following is determined as $\|v^k - v^{k-1}\|_0 / \|v^{k-1}\|_0 < \epsilon$, where $\epsilon = 10^{-5}$. We remark that the tolerance was chosen such that in the refinement studies, where convergence of DD was observed, the approximation error in the $H^1 \times H^1$ norm decreased with the expected order as $O(h)$ (we recall that $P_1 - P_1$ elements were used).

To test robustness of the DD solver we consider the setup of the single cell experiment used with the monolithic approaches in Section 6.2. Here the Robin problem (i) and the Neumann problem (ii) shall be solved by LU to eliminate effects of inexact subdomain solvers. We remark that due to the conforming triangulation, see Figure 5.1, and P_1 discretization of both V_i and V_e the discrete transmembrane potential is computed simply by interpolation. Figure 6.1 plots evolution of the potential difference $\|v_h^k - v_h^{k-1}\|_0 / \|v_h^{k-1}\|_0$ with DD iterations for several mesh resolutions and time step values $\Delta t \leq 1$. It can be seen that for $\Delta t = 1$ the algorithm converges in about 5 iterations irrespective of the spatial discretization. For smaller timesteps convergence is delayed on finer meshes and for $\Delta t = 10^{-4}$ the iterations diverge.

Considering the results in Figure 6.1 we conclude that in the form presented here domain decomposition is not a robust algorithm for the single-dimensional primal formulation of the EMI equations. However, due to its simplicity (relative to e.g. the multi-dimensional formulations) and speed (see Section 6.4), DD might be the method of choice in practical cardiac modeling. It is therefore natural to ask whether the divergence conditions are likely to arise in real applications. To address this question we perform a scaling analysis based on the membrane dynamical condition $C_m \frac{\partial v}{\partial t} \sim \sigma_i \nabla u_i \cdot \boldsymbol{n}_i$. Letting L be a characteristic length scale and $\nabla' = L\nabla$ the equation becomes $(LC_m/\sigma_i)\frac{\partial v}{\partial t} = \nabla' u_i \cdot \boldsymbol{n}_i$ and the pre-factor $T = LC_m/\sigma_i$ can be seen to have the unit of seconds. Following (22) let us insert $C_m = 1\mu F/cm^2$, $\sigma_i = 10mS/cm$, $L = 100\mu m$, where the length scale is determined by the size of a typical cell used by the authors'. We remark that this a sensible choice given the setup of our experiments. As a result $T = 10^{-6}s$ and thus $\Delta t = 1$ in the experiments reported here corresponds to a time step of 10^{-6} seconds. Note that this is the finest time step considered in (11). Furthermore, therein the spatial resolution is $2\mu m$ while here with $n_h = 2^8$ and $L = 100\mu m$ the mesh size is approximately $0.4\mu m$. In summary, the conditions for divergence of the DD algorithm discussed here can be encountered not far away from the parameter regime in (11).

6.4 Solver Comparison

To complete our discussion of EMI solvers we finally address the speed of the different algorithms. Recall that until this point results for all solvers, but the monolithic single-dimensional primal one, were obtained using LU in the construction. Such an implementation, however, is not scalable. Here we show that the proposed algorithms are order optimal if LU is replaced by suitable multilevel methods.

Referring to the legend of Figure 6.2 we shall compare 8 different methods. Single-dimensional primal formulation (sp) shall be solved either with the BoomerAMG($\mathcal{A}$)-preconditioner (mono), or diagonal preconditioner (6.7). In the latter case all the blocks of the diagonal preconditioner are approximated by single algebraic multigrid V-cycle. The single-dimensional mixed (sm) formulations shall use the $\mathcal{B}_1$ preconditioner (6.10) with the leading block approximated by $\boldsymbol{H}$(div) multigrid HypreAMS of (14). We recall that the solver in general is not independent of the discretization, cf. Table 6.3, however, HypreAMS does not work well for the robust preconditioner $\mathcal{B}_2$. Finally, the multi-dimensional primal (mp) and multi-dimensional mixed (mm) formulations shall use (6.15) and (6.16) with the Ω_i, Ω_e and Ω subproblems of the preconditioners approximated by multigrid (BoomerAMG in case of (6.15) and HypreAMS for (6.16)). The fractional operators will be computed exactly. In addition, two implementations of Robin/Neumann domain decomposition algorithm are considered with the subproblems solved either exactly by LU or by 4 V-cycles of BoomerAMG. Finally, a reference solution time is provided by timings of exact solution of the single-dimensional formulation (sp-LU).

We compare the algorithms using the single cell setup of the previous experiments with $\Delta t = 10^{-3}$ where this value was chosen with the intention to not favor any of the methods. Using tolerance $\epsilon = 10^{-3}$ for DD and 10^{-12} for CG/MinRes, Figure 6.2 reports solution times[3] of the algorithms for $2^3 \leq n_h \leq 2^{10}$. It can be seen that with the exception of LU-based solvers all the methods are indeed order optimal. The monolithic approach (sp-mono) and the multi-dimensional primal (mp) solver are then the fastest solvers while solvers for the mixed formulations (sm-$\mathcal{B}_1$, mm) are the slowest. For $n_h = 2^{10}$ the solution times of the solvers were cca. 26s, 140s, 640s, 570s respectively. We remark that the mixed formulations have approximately 5 times more unknowns compared to the single-dimensional ones. Thus the cost per degree of freedom is comparable between the two approaches. Considering the scalable DD implementation (DD-AMG) the timing for $n_h = 2^{10}$ is cca. 160s. The method thus has a similar cost to multi-dimensional primal formulation.

---

[3] The timings include setup times of the preconditioners for monolithic methods. For DD the subproblems were assembled and their solvers constructed only once.

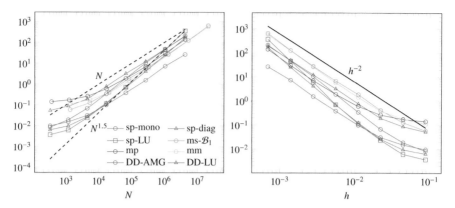

Fig. 6.2: Solution times (including preconditioner setup) of the solvers for the EMI system (5.1) in terms of number of unknowns N or mesh size h. Single cell model with $\Delta t = 10^{-3}$ is used. All but multi-dimensional primal solver with LU (sp-LU) and LU-based domain decomposition solver (DD-LU) are order optimal. The former scales as $N^{3/2}$. Legend is shared between the figures.

References

1. Bærland T, Kuchta M, Mardal KA (2019) Multigrid methods for discrete fractional Sobolev spaces. SIAM Journal on Scientific Computing 41(2):A948–A972
2. Balay S, Abhyankar S, Adams MF, Brown J, Brune P, Buschelman K, Dalcin L, Dener A, Eijkhout V, Gropp WD, Karpeyev D, Kaushik D, Knepley MG, May DA, McInnes LC, Mills RT, Munson T, Rupp K, Sanan P, Smith BF, Zampini S, Zhang H, Zhang H (2019) PETSc Web page. https://www.mcs.anl.gov/petsc, URL https://www.mcs.anl.gov/petsc
3. Bonito A, Pasciak J (2015) Numerical approximation of fractional powers of elliptic operators. Mathematics of Computation 84(295):2083–2110
4. Braess D (2007) Finite Elements: Theory, Fast Solvers, and Applications in Solid Mechanics. Cambridge University Press, Cambridge
5. Bramble J, Pasciak J, Vassilevski P (2000) Computational scales of Sobolev norms with application to preconditioning. Mathematics of Computation 69(230):463–480
6. Brenner S, Scott R (2007) The mathematical theory of finite element methods, vol 15. Springer Science & Business Media
7. Brezzi F (1974) On the existence, uniqueness and approximation of saddle-point problems arising from Lagrangian multipliers. Revue française d'automatique, informatique, recherche opérationnelle Analyse numérique 8(2):129–151
8. Discacciati M, Quarteroni A, Valli A (2007) Robin–Robin domain decomposition methods for the Stokes–Darcy coupling. SIAM Journal on Numerical Analysis 45(3):1246–1268
9. Girault V, Raviart PA (2012) Finite Element Methods for Navier-Stokes Equations: Theory and Algorithms, vol 5. Springer Berlin Heidelberg, Berlin, Heidelberg
10. Hernandez V, Roman JE, Vidal V (2005) SLEPc: A scalable and flexible toolkit for the solution of eigenvalue problems. ACM Trans Math Software 31(3):351–362
11. Jæger KH, Tveito A (2020) Efficient numerical solution of the EMI model representing extracellular space (E), the cell membrane (M) and the intracellular space (I) of a collection of cardiac cells. Preprint
12. Jæger KH, Hustad KG, Cai X, Tveito A (2020) Operator splitting and finite difference schemes for solving the EMI model. In: Tveito A, Mardal KA, Rognes ME (eds) Modeling excitable tissue - The EMI framework, Simula Springer Notes in Computing, SpringerNature
13. Klawonn A, Widlund OB, Dryja M (2002) Dual-primal FETI methods for three-dimensional elliptic problems with heterogeneous coefficients. SIAM Journal on Numerical Analysis 40(1):159–179
14. Kolev TV, Vassilevski PS (2012) Parallel auxiliary space AMG solver for H(div) problems. SIAM Journal on Scientific Computing 34(6):A3079–A3098
15. Kuchta M, Mardal KA (2020) Software for EMI - Iterative solvers for EMI models. URL https://doi.org/10.5281/zenodo.3771212

16. Kuchta M, Nordaas M, Verschaeve JCG, Mortensen M, Mardal KA (2016) Preconditioners for saddle point systems with trace constraints coupling 2D and 1D domains. SIAM Journal on Scientific Computing 38(6):B962–B987
17. Kuchta M, Mardal KA, Rognes ME (2020) Solving the EMI equations using finite element methods. In: Tveito A, Mardal KA, Rognes ME (eds) Modeling excitable tissue - The EMI framework, Simula Springer Notes in Computing, SpringerNature
18. Mandel J, Dohrmann CR, Tezaur R (2005) An algebraic theory for primal and dual substructuring methods by constraints. Applied Nnumerical Mmathematics 54(2):167–193
19. Mardal KA, Winther R (2011) Preconditioning discretizations of systems of partial differential equations. Numerical Linear Algebra with Applications 18(1):1–40
20. Nielsen BF, Mardal KA (2013) Analysis of the minimal residual method applied to ill posed optimality systems. SIAM Journal on Scientific Computing 35(2):A785–A814
21. Smith B, Bjorstad P, Gropp W (2004) Domain Decomposition: Parallel Multilevel Methods for Elliptic Partial Differential Equations. Cambridge University Press, Cambridge
22. Tveito A, Jæger KH, Kuchta M, Mardal KA, Rognes ME (2017) A cell-based framework for numerical modeling of electrical conduction in cardiac tissue. Frontiers in Physics 5:48
23. Yang UM, et al. (2002) Boomeramg: a parallel algebraic multigrid solver and preconditioner. Applied Numerical Mathematics 41(1):155–177
24. Zampini S (2016) PCBDDC: a class of robust dual-primal methods in PETSc. SIAM Journal on Scientific Computing 38(5):S282–S306
25. Zampini S, Tu X (2017) Multilevel balancing domain decomposition by constraints deluxe algorithms with adaptive coarse spaces for flow in porous media. SIAM Journal on Scientific Computing 39(4):A1389–A1415

Chapter 7
Improving Neural Simulations with the EMI Model

Alessio Paolo Buccino[1,2], Miroslav Kuchta[3], Jakob Schreiner[3], and Kent-André Mardal[3,4]

Abstract Mathematical modeling of neurons is an essential tool to investigate neuronal activity alongside with experimental approaches. However, the conventional modeling framework to simulate neuronal dynamics and extracellular potentials makes several assumptions that might need to be revisited for some applications. In this chapter we apply the EMI model to investigate the ephaptic effect and the effect of the extracellular probes on the measured potential. Finally, we introduce reduced EMI models, which provide a more computationally efficient framework for simulating neurons with complex morphologies.

7.1 Introduction

In recent years, huge efforts and resources have been spent in computational modeling of neuronal activity. For example, the Blue Brain Project (18; 16)(`https://bbp.epfl.ch/nmc-portal/welcome`) has constructed and distributed several hundreds of biophysically detailed cell models (*multi-compartment models*) from rat sensory cortex. A similar effort is being conducted by the Allen Institute of Brain Science, whose cell-type database (8) (`https://celltypes.brain-map.org/`) includes hundreds of cell models both from mice and even from humans. As the experimental data used become more comprehensive and available, these models are expected to become elaborated and more accurate in reproducing neuronal dynamics. However, the modeling framework which is commonly used to simulate these

---

[1]Bio Engineering Laboratory, Department of Biosystems and Science Engineering, ETH Zurich, Switzerland
[2]Center for Integrative Neuroplasticity (CINPLA), Faculty of Mathematics and Natural Sciences, University of Oslo, Oslo, Norway
[3]Simula Research Laboratory, Norway
[4]Department of Mathematics, University of Oslo, Oslo, Norway

The Author(s) 2021
A. Tveito et al. (eds.), *Modeling Excitable Tissue*, Simula SpringerBriefs on Computing 7, https://doi.org/10.1007/978-3-030-61157-6_7

multi-compartment models makes several key assumptions that might be violated for certain applications.

The most widely used approach to simulate neuronal dynamics of neurons is the *cable equation*. The solution of this equation enables one to compute transmembrane currents for each of the model compartment. In order to simulate extracellular potentials, we can use *volume conduction theory* and sum the individual contributions of the currents to the electric potential at any point in space (6). Whereas the use of this modeling framework has been the gold standard to simulate neuronal activity for decades (19; 6), there are some important assumptions that need to be discussed:

- **A neuron is represented as a discrete set of nodes.** Multi-compartment models split the neuronal morphologies into a discrete set of segments. Therefore, neurons are not represented as a *continuum* and this might affect the accuracy of the simulations. However, this assumption can be alleviated by using very small segments that can accurately replicate the neuronal complex geometry.

- **Extracellular potentials are assumed to be constant.** When solving the cable equation, the extracellular potentials outside the membrane are assumed to be constant. This assumption is harder to relax, as it prevents to include so-called *ephaptic* coupling in the simulation (9; 1). Ephaptic coupling refers to the effect of extracellular potentials on the neuronal dynamics. The use of the EMI model allows one to include these phenomena in the simulation, both to simulate the effect other neurons' activity or the same neuron's activity (self-ephaptic) has on the membrane potential.

- **The extracellular space is assumed to be homogeneous.** The most common approach to compute extracellular potentials arising from neuronal currents is to use volume conduction theory with the point-source or line-source approximations (6), which assume that the extracellular medium is homogeneous (in addition to linear, isotropic, and infinite). However, this assumption is clearly violated when using extracellular devices to record neuronal activity, which introduce a clear inhomogeneity in the extracellular space. Extracellular probes can be explicitly modeled in the extracellular space with the EMI model and they show to greatly contribute to the recorded signals (3).

In applications where any of the assumptions listed above may be violated the EMI model (10, (1.30)) provides a suitable modeling framework. In particular, the geometry of the neuron (and the extracellular space) is accurately represented. Here we will show how the model is convenient to study both the ephaptic coupling of neurons (Section 7.3) and the effects of extracellular probes on the recorded electric potentials (Section 7.4). However, the detailed representation of the geometries makes EMI much more computationally intense than the standard modeling framework (21; 3). This limits the complexity of the simulation mainly to simple neuronal morphologies, such as ball-and-stick neurons (3). In order to target more realistic morphologies Section 7.5 discusses the reduced EMI model where the (three-dimensional) intracellular space is lumped to a curve that is the centerline of the neuron.

7.2 EMI Simulations of Neurons using the neuronmi Python Package

Before discussing the applications let us first briefly introduce a Python package called neuronmi[1], which has been used for the simulation results reported further. neuronmi provides a high-level application programming interface (API) to enable users to easily set up and run EMI simulations of neurons.

The workflow of the neuronmi package consists of two parts. First, a mesh needs to be generated. This is done with the generate_mesh function, that uses gmsh (7) as backend. With this function the user can choose different kinds of neurons to place in the mesh and optionally place a probe device in the extracellular space. Mesh resolution and sizes of the bounding box can also be adjusted, as well as parameters of the neurons and the probe. In the following code example, we create a mesh with a ball–and–stick neuron (bas) and a microwire probe. By default, the center of the soma is at $(0,0,0)\,\mu m$, the dendrite extends in the positive z direction and the axon in the negative z direction.

```
import neuronmi
mesh_folder = neuronmi.generate_mesh(neurons='bas',
                                     probe='microwire')
```

Once a mesh is generated, the EMI simulation can be invoked with the simulate_emi function, which implements the finite element method for the multi-dimensional mixed formulation (14, (5.14)) following the discretization proposed in (20). Through a set of parameters, the user can stimulate the neuron with a synaptic input, a step current, or a pulse current. Alternatively, the probe contacts can be used to stimulate the neuron extracellularly. By default, the neuron receives a synaptic input on its dendrite. The user can probe electric potentials u at any point in the mesh, while transmembrane currents i and membrane potentials v are available at facets on the neuron surface. The full solutions are also saved as pvd or xdmf files. The simulation is run as follows:

```
u, i, v = neuronmi.simulate_emi(mesh_folder,
                                u_probe_locations=points_v,
                                i_probe_locations=points_i,
                                v_probe_locations=points_v)
```

---

[1] https://github.com/MiroK/nEuronMI

Several parameters can be set to customize the mesh generation and the simulation. For further details, we refer to the package documentation (`https://neuronmi.readthedocs.io/en/latest/`).

7.3 Investigating the Ephaptic Effect between Neurons

Neurons are surrounded by the electrically conductive extracellular space. Groups of neurons create fluctuations in the local extracellular electrical field. These fluctuations in turn influence the intracellular electrical field through the *ephaptic* effect. Ephaptic coupling cannot influence neurons at rest, however, it can affect the spike timings of a neuron receiving suprathreshold stimulus.

We will illustrate how the EMI model can be used to compute the ephaptic coupling between two ball–and–stick neurons embedded in an extracellular space. The simulation is based on the `neuronmi` package detailed above. One of the neurons is stimulated with a synaptic input which elicits an action potential. The intracellular potential in the other neuron is sampled. We ran several experiments increasing the distance between the neurons.

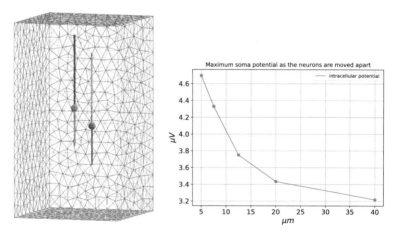

Fig. 7.1: Two ball–and–stick neurons embedded in an extracellular space (left) and the increase in the intracellular potential due to ephaptic coupling (right).

The intracellular potential is sampled in the centre of the ephaptically stimulated neuron (right) to measure the strength of the synaptic coupling. The deflection is 4.7 μV when the neurons are 5 μm apart and decreases to 3.2 μV when they are 40 μm apart with the soma of the stimulated neuron adjacent to the axon of the other neuron.

While in these simulations we only showed the effect of a single spike on an adjacent neuron, the occurrence of synchronous activity in populations of neurons can cause larger degrees of ephaptic coupling. The neuronmi package enables users to instantiate several neurons in the mesh and to define their inputs (which can be synchronous), hence allowing in principle to investigate ephaptic effects at the population level.

7.4 Investigating the Effect of Measuring Devices on Extracellular Potentials

While the presence of recording devices is usually ignored in the computation of extracellular potentials, recent findings suggest that newly developed silicon-based devices, or Multi-Electrode Arrays (MEAs), have a strong effect on the measured signals (3).

Using neuronmi, which is built on the EMI model, one can easily incorporate the neural devices in the mesh and investigate their effects on the recorded signals. To demonstrate this, we built meshes with a simple ball-and-stick neuron and different types of neural probes in its vicinity:

Microwire: the first type of probe represents a microwire. For this kind of probes we used a cylindrical insulated model with 30 μm diameter. The extracellular potential, after the simulations, was measured at the center of tip of the cylinder. The microwire probe is shown in Figure 7.2A alongside with the simplified neuron.

Neuronexus (MEA): the second type of probe model represents a commercially available silicon MEA (A1x32-Poly3-5mm-25s-177-CM32 probe from Neuronexus Technologies), which has 32 electrodes in three columns (the central column has 12 recording sites and first and third columns have 10) with hexagonal arrangement, a y-pitch of 18 μm, and a z-pitch of 22 μm. The electrode radius is 7.5 μm. This probe has a thickness of 15 μm and a maximum width of 114 μm, and it is shown in Figure 7.2B.

Neuropixels (MEA): the third type of probe model represents the Neuropixels silicon MEA (11). The original probe has more than 900 electrodes over a 1 cm shank. The probbe is 70 μm wide and 20 μm thick. In our mesh, shown in Figure 7.2C we used 24 12x12 μm recording sites arranged in the *chessboard* configuration with an inter-electrode-distance of 25 μm (11).

We ran EMI simulations using the meshes with and without the probe in the extracellular space (Figure 7.2A-B-C show the meshes with the probe), and we compared the obtained extracellular action potentials - EAPs (Figure 7.2D-E-F, blue without

probe - orange with probe). The probes were placed in the extracellular space at a
distance of 30 µm from the center of the neuron.

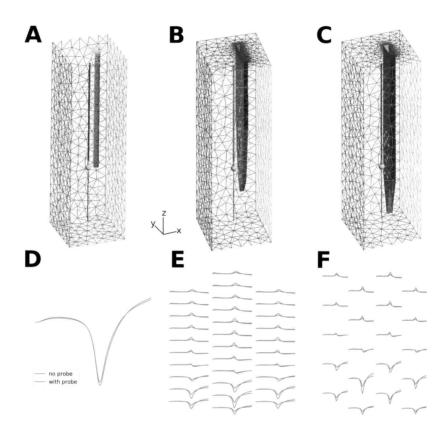

Fig. 7.2: Effect of different probes on the recorded potentials. (A-B-C) Meshes
including a neuron and a microwire (A), a neuron and a Neuronexus probe, a neuron
and a Neuropixels probe (C). (D-E-F) Extracellular action potentials computed at the
electrodes' location without (blue) and with the probe (orange) in the extracellular
space. Large MEAs seem to strongly affect the recorded signals (E-F).

Microwire probes do not affect the recorded potentials, with an EAP peak of
$-21.63\,\mu V$ without the probe and of $-20.53\,\mu V$ with the probe (Figure 7.2D). However, when recording with silicon MEAs, the extracellular potentials are strongly
affected. For the Neuronexus probe (Figure 7.2D), the EAP peak without the probe
is $-30.47\,\mu V$, while with the probe it is $-56.09\,\mu V$ (peak ratio=1.84). For the Neuropixels probe (Figure 7.2E), the EAP peak without the probe is $-32.73\,\mu V$, while
with the probe it is $-63.63\,\mu V$ (peak ratio=1.94). The probe effect is probably due to

the insulating properties of the silicon probes, which act as *electrical walls* for the generated currents. For further details and analysis we refer to (3).

7.5 Reduced EMI Model

In the examples considered thus far the problem geometry was simple allowing for computations on a serial desktop computer. In order to apply the EMI framework to realistic neurons two challenges need to be addressed: representation of neurons in the form of a finite element mesh and efficient solvers for large linear systems due to the EMI equations. However, even with the order optimal algorithms discussed in (12, Chapter 6) and efficient mesh generators for neuron surface geometries, see e.g. (17), the computational cost of the $3D$-$3D$ EMI models remains large (compared to the conventional approaches). As a computationally feasible alternative we shall next discuss the $3D$-$1D$ models.

Topological order reduction is a modeling technique used e.g. in reservoir simulations (4) or studies of tissue perfusion (5), which exploits geometrical properties of the system in order to derive its reduced model. Viewing a dendrite (branch) as generalized cylinder with length L and radius R we observe that $R \ll L$ and that in a typical domain of interest the neuron's volume is negligible compared to its surroundings. This property motivates a reduced representation of the neuron in terms of a (one dimensional) curve, the centerline, along with a function R, which provides radius of the crossection at each point of the line. An illustration of the concept can be seen in Figure 7.3. Thus, referring to the notation of Chapter 5, Ω_i is reduced to a line while Ω_e newly occupies the entire domain, i.e. $\Omega = \Omega_e$. In turn, the reduced EMI model presents a coupling between three dimensional extracellular space and the one-dimensional intracellular space. We remark that the membrane is one-dimensional as well.

In order to apply order reduction to the EMI model we consider the single-dimensional primal formulation (5.6). Note that therein, the coupling on the membrane Γ requires that both u_e and u_i are restricted from Ω_e and Ω_i respectively by the dedicated trace operator. In a reduced model $\Omega_i = \Gamma$, $\Gamma = \Lambda$ and thus u_i needs not to be restricted. On the other hand, restriction of u_e to *curve* Λ can no longer be realized as a trace since such an operation is not well-defined for H^1 functions, see e.g. (5). To define a value of the extracellular potential on Λ let us introduce an averaging operator

$$\Pi u(x) = \bar{u}(x) = |C_R(x)|^{-1} \int_{C_R(x)} u(y) \, dS, \quad x \in \Gamma, u \in H^1(\Omega).$$

Here $C_R(x) = \{y \in \Gamma; (y - x) \perp n(x)\}$ with $n(x)$ being the unit tangent vector of Λ at x, cf. Figure 7.3. Thus, Πu_e is computed by sampling u_e on the original (two-

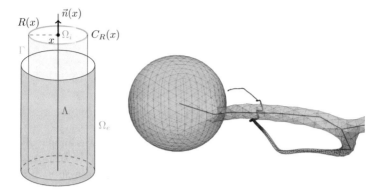

Fig. 7.3: Reduced representation of neurons. (Left) Neuron Ω_i is collapsed to centerline Λ. Extracellular potential defined on Ω is evaluated at Λ by averaging u_e on the cylindrical surface Γ. (Right) Surface mesh and centerline representation of RatS1-6-39 neuron generated by AnaMorph (17). The dendritic part satisfies $R \ll L$ property, while the reduced model assumptions do not hold on the soma.

dimensional) neuron surface. However, in practical computations we assume that Γ has a circular cross section so that $|C_R(x)| = 2\pi R(x)$.

Using Π the reduced single-scale primal formulation of the EMI model reads: Find $u_e \in H^1_{\Gamma^D_e}(\Omega)$, $u_i \in H^1(\Lambda)$ such that for all $v_e \in H^1_{\Gamma^D_e}(\Omega)$, $v_i \in H^1(\Lambda)$

$$\int_\Omega \sigma_e \nabla u_e \cdot \nabla v_e \, dx + \int_\Lambda 2\pi R \frac{C_m}{\Delta t} \bar{u}_e \bar{v}_e \, ds - 2\pi R \int_\Lambda \frac{C_m}{\Delta t} \bar{v}_e u_i \, ds$$

$$= \int_\Lambda 2\pi R f \bar{v}_e \, ds,$$

$$-\int_\Lambda 2\pi R \frac{C_m}{\Delta t} v_i \bar{u}_e \, ds + \int_\Lambda \pi R^2 \sigma_i \nabla u_i \cdot \nabla v_i \, ds + \int_\Lambda 2\pi R \frac{C_m}{\Delta t} u_i v_i \, ds$$

$$= -\int_\Lambda 2\pi R f v_i \, ds.$$

(7.1)

Here, the factors $2\pi R$ arise in reducing the integration domain from Γ to Λ and similarly for πR^2 and Ω_i. Thus, defining the reduced specific membrane capacitance $C_m = 2\pi R C_m$ and the reduced intracellular conductivity $\sigma_i = \pi R^2 \sigma_i$ the operator form of (7.1) can be seen to be (6.2) with the new restriction operators $T_i = I$ and $T_e = \Pi$. Note also that in (7.1) the function f, which characterizes the membrane dynamics, is defined on Γ.

For the proof of well-posedness of (7.1) as well as a detailed discussion of modeling assumptions, which allow for model reduction from (5.6) the reader is referred to (4). Moreover the reduced multi-dimensional primal formulation (5.8) is analyzed in (13).

To assess the reduced model (7.1) the surface mesh of a rat neocortex neuron RatS1-6-39 from the NeuroMorpho database (2) has been generated using AnaMorph (17). The neuron has been embedded into a (box-shaped) domain Ω such that $|\Omega|/|\Omega_i| \sim 100$ with the resulting geometry meshed by gmsh. The full $3D$-$3D$ single-dimensional primal formation has been used to compute the response to a 1 ms synaptic stimulus of 50 nA. Referring to Figure 7.4 the lower branch close to node number 4 has been stimulated. Using the centerline representation of the neuron the response has also been computed with (7.1). We remark that P_1-P_1 elements were used with both formulations and that spaces were setup on conforming meshes. In particular, with (7.1) the discretization of Λ consisted of the edges of the cells of Ω. However, such a geometrically conforming discretization is not required in the reduced model. In fact, the meshes of Λ and Ω can be independent, see (4). The reduced model then resulted in 4587 unknowns, which is to be contrasted with 18248 unknowns due to (5.6). In turn, the simulation time using the reduced model is about 110 seconds while the full EMI model required cca. 340 seconds to complete.

The two models are compared in Figure 7.4 which shows values of the computed intracellular potentials at different points along the centerline. In general, there is qualitative agreement between the model predictions. However, the reduced model can be seen to tend to underestimate both the minima and the maxima, while the excitation occurs faster compared to the full model. More precisely, the peak potentials due to the full model at points 2-5 were $\{27.64, 26.09, 22.87, 12.64\}$ mV with occurrences after $\{2.08, 2.11, 2.15, 2.57\}$ ms. For the reduced model the maxima $\{22.31, 21.65, 19.39, 16.02\}$ mV were recorded at $\{1.56, 1.58, 1.51, 1.86\}$ ms. In addition to the intracellular potentials, the extracellular potentials were compared by sampling $6.52\mu m$ away from the soma center (node 1 in Figure 7.4). We remark that the soma radius was $5.71\mu m$. It can be seen that with $-2.99\mu V < u_e < 1.07\mu V$ for (7.1) and $-1.99\mu V < u_e < 0.67\mu V$ for (5.6) the reduced model overestimates the extrema. As with the extracellular potentials there is a temporal shift in the response; the negative peak is observed at 1.26ms, respectively 1.87ms.

While results of the comparison in Figure 7.4 shall be viewed as preliminary we argue that they illustrate sufficiently the potential of reduced EMI models. In particular, the reduced model is able to capture qualitatively the properties of the full EMI simulations. However, clear differences, especially in the temporal shifts of the peaks, have been observed. In the future we aim to investigate if suitable scaling of the stimulus and/or the parameters of the membrane ODEs can reduce the prediction error. In addition, the modeling error of the reduced model shall be evaluated similar to (15). More specifically, the soma, being approximated as a sphere in the $3D$-$3D$ model, cannot be represented as a slender cylinder (unlike the dendrites and axons). Thus the model reduction assumptions are not met on the soma. While localized in space, the effect of this error on temporal predictions should be analyzed. In turn, improved reduced models, which take into account the spherical nature of the soma, e.g. in construction of averaging operators, might be needed.

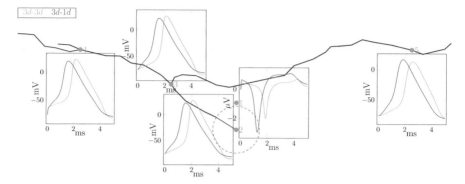

Fig. 7.4: Comparison of full $3D$-$3D$ and reduced $3D$-$1D$ EMI models. RatS1-6-39 neuron is stimulated close to node 4 in the dendritic part (plotted in blue). Intracellular potentials (nodes 2-5) are measured on the centerline with node 2 being the soma (in red, dashed line indicates the radius) center. Extracellular potentials are compared in node 1 next to soma. Axis of each response plot is anchored at the center next to the measurement point. The full and reduced EMI models provide qualitatively similar predictions.

7.6 Conclusions

In this chapter, we have showcased some applications in which the EMI model can be a viable alternative to standard modeling frameworks in order to investigate aspects of neuronal activity and recorded signals. We introduced an open-source software package named **neuronmi** to easily assemble meshes including simplified neurons and probes in the extracellular space. Finally, we performed preliminary investigations into accuracy and solution cost of a reduced $3D$-$1D$ EMI model suitable for simulating complex neuronal morphologies.

References

1. Anastassiou CA, Perin R, Markram H, Koch C (2011) Ephaptic coupling of cortical neurons. Nature Neuroscience 14(2):217
2. Ascoli GA, Donohue DE, Halavi M (2007) Neuromorpho.org: a central resource for neuronal morphologies. Journal of Neuroscience 27(35):9247–9251
3. Buccino AP, Kuchta M, Jæger KH, Ness TV, Berthet P, Mardal KA, Cauwenberghs G, Tveito A (2019) How does the presence of neural probes affect extracellular potentials? Journal of Neural Engineering 16(2):026030
4. Cerroni D, Laurino F, Zunino P (2019) Mathematical analysis, finite element approximation and numerical solvers for the interaction of 3d reservoirs with 1d wells. GEM-International Journal on Geomathematics 10(1):4
5. D'Angelo C, Quarteroni A (2008) On the coupling of 1d and 3d diffusion-reaction equations: application to tissue perfusion problems. Mathematical Models and Methods in Applied Sciences 18(08):1481–1504
6. Einevoll GT, Kayser C, Logothetis NK, Panzeri S (2013) Modelling and analysis of local field potentials for studying the function of cortical circuits. Nature Reviews Neuroscience 14(11):770–785
7. Geuzaine C, Remacle JF (2009) Gmsh: A 3-d finite element mesh generator with built-in pre- and post-processing facilities. International Journal for Numerical Methods in Engineering 79(11):1309–1331
8. Gouwens NW, Berg J, Feng D, Sorensen SA, Zeng H, Hawrylycz MJ, Koch C, Arkhipov A (2018) Systematic generation of biophysically detailed models for diverse cortical neuron types. Nature Communications 9(1):710
9. Holt GR, Koch C (1999) Electrical interactions via the extracellular potential near cell bodies. Journal of Computational Neuroscience 6(2):169–184
10. Jæger KH, Tveito A (2020) Derivation of a cell-based mathematical model of excitable cells. In: Tveito A, Mardal KA, Rognes ME (eds) Modeling excitable tissue - The EMI framework, Simula Springer Notes in Computing, SpringerNature
11. Jun JJ, Steinmetz NA, Siegle JH, Denman DJ, Bauza M, Barbarits B, Lee AK, Anastassiou CA, Andrei A, Aydın Ç, et al. (2017) Fully integrated silicon probes for high-density recording of neural activity. Nature 551(7679):232–236
12. Kuchta M, Mardal KA (2020) Iterative solvers for cell-based EMI models. In: Tveito A, Mardal KA, Rognes ME (eds) Modeling excitable tissue - The EMI framework, Simula Springer Notes in Computing, SpringerNature
13. Kuchta M, Laurino F, Mardal KA, Zunino P (2020) Analysis and approximation of mixed-dimensional pdes on 3d-1d domains coupled with lagrange multipliers. arXiv preprint arXiv:200402722

14. Kuchta M, Mardal KA, Rognes ME (2020) Solving the EMI equations using finite element methods. In: Tveito A, Mardal KA, Rognes ME (eds) Modeling excitable tissue - The EMI framework, Simula Springer Notes in Computing, SpringerNature
15. Laurino F, Zunino P (2019) Derivation and analysis of coupled PDEs on manifolds with high dimensionality gap arising from topological model reduction. ESAIM: M2AN 53(6):2047–2080
16. Markram H, et al. (2015) Reconstruction and simulation of neocortical microcircuitry. Cell 163(2):456–492
17. Mörschel K, Breit M, Queisser G (2017) Generating neuron geometries for detailed three-dimensional simulations using AnaMorph. Neuroinformatics 15(3):247–269
18. Ramaswamy S, et al. (2015) The neocortical microcircuit collaboration portal: a resource for rat somatosensory cortex. Frontiers in Neural Circuits 9
19. Sterratt D, Graham B, Gillies A, Willshaw D (2011) Principles of computational modelling in neuroscience. Cambridge University Press
20. Tveito A, Jæger KH, Kuchta M, Mardal KA, Rognes ME (2017) A cell-based framework for numerical modeling of electrical conduction in cardiac tissue. Frontiers in Physics 5:48
21. Tveito A, Jæger KH, Lines GT, Paszkowski Ł, Sundnes J, Edwards AG, Māki-Marttunen T, Halnes G, Einevoll GT (2017) An evaluation of the accuracy of classical models for computing the membrane potential and extracellular potential for neurons. Frontiers in Computational Neuroscience 11:27

Index

The Author(s) 2021
A. Tveito et al. (eds.), *Modeling Excitable Tissue*, Simula SpringerBriefs
on Computing 7, https://doi.org/10.1007/978-3-030-61157-6

Printed in the United States
By Bookmasters